GREAT
  in Science and Technology

GREAT MARITIME ACHIEVERS
IN SCIENCE AND TECHNOLOGY

George MacBeath

CHARLES P. ALLEN HIGH
196 ROCKY LAKE DR.
BEDFORD, N.S. B4A 2T6

Copyright © George MacBeath, 2004.

All rights reserved. No part of this work may be reproduced or used in any form or by any means, electronic or mechanical, including photocopying, recording, or any retrieval system, without the prior written permission of the publisher or a licence from the Canadian Copyright Licensing Agency (Access Copyright). To contact Access Copyright, visit www.accesscopyright.ca or call 1-800-893-5777.

Edited by Sabine Campbell and Laurel Boone.
Cover photos: Trilobite model, New Brunswick Museum; *Silver Dart*, National Archives of Canada PA-061741; kerosene lamp, King's Landing Historical Settlement; nursing sister and soldiers, 1900, National Archives of Canada C-051799; W.R. Turnbull, Canada Aviation Museum. Reproduced by permission.

Cover design by Paul Vienneau.
Book design by Julie Scriver.
Printed in Canada by Transcontinental.
10 9 8 7 6 5 4 3 2 1

Library and Archives Canada Cataloguing in Publication

MacBeath, George, 1924-
Great maritime achievers in science and technology / George MacBeath.

Includes index.
ISBN 0-86492-380-5

1. Scientists — Maritime Provinces — Biography.
2. Inventors — Maritime Provinces — Biography. I. Title.

Q141.M14 2004   509.2'2715   C2004-905123-7

Published with the financial support of the Canada Council for the Arts, the Government of Canada through the Book Publishing Industry Development Program, and the New Brunswick Culture and Sports Secretariat.

Goose Lane Editions
469 King Street
Fredericton, New Brunswick
CANADA E3B 1E5
www.gooselane.com

CONTENTS

7	Foreword
11	Abraham Gesner
15	Robert Foulis
19	Titus Smith
23	Moses Henry Perley
25	Andrew Downs
29	Benjamin Tibbets
33	Charles Fenerty
37	Thomas Hall
41	Francis Peabody Sharp
45	Frederick Newton Gisborne
49	Simon Newcomb
53	John William Dawson
57	David Honeyman
59	James Robb
63	William Brydone Jack
67	George Frederic Matthew
71	George Mercer Dawson
75	Loring Woart Bailey
79	William Diller Matthew

81	Charles Frederick Hartt
83	William Francis Ganong
87	Henry G.C. Ketchum
91	Grace Annie Lockhart
93	William MacIntosh
95	John Clarence Webster
99	Alexander Graham Bell, Mabel Hubbard Bell, J.A.D. McCurdy, and Frederick "Casey" Baldwin
105	Wallace Rupert Turnbull
109	Georgina Fane Pope
113	Margaret MacDonald
115	Robie Tufts
118	Illustration Credits
119	Acknowledgements
121	Index

FOREWORD

Great Maritime Achievers in Science and Technology is the result of a number of ideas that joined together and kept getting bigger and bigger. The concept originated with the board of directors of Science East, New Brunswick's hands-on science centre, in Fredericton. At first, we thought of creating a hall of honour at Science East, similar to the one at the National Museum of Science and Technology, to pay tribute to New Brunswickers who had made outstanding contributions to the fields of science and technology. Further enthusiastic suggestions followed. Why not make this a travelling exhibition, with which Science East had plenty of experience? Then the Museums Assistance Program of Heritage Canada urged that the project be expanded to include all three of the Maritime Provinces. And as the project grew, Goose Lane Editions envisioned an illustrated book to complement the touring exhibition.

As research progressed, compelling stories of curiosity and imagination emerged, mind-stretching experiences that led to discovery and unfolded mysteries. The travelling exhibition includes major presentations featuring achievements of about a dozen people and smaller displays showcasing the work of around twenty others. The book captures and expands on their feats. This is the story of invention, from the early days to the triumph of discovery. But it is also the story of people, the behind-the-scenes personal stories and details, the work, dreams, discouragements, and triumphs of these Maritime Achievers. Their accomplishments are part of our rich heritage as Maritimers and as Canadians.

In heading up the research phase of this project, Dr. George MacBeath called on two colleagues to help with the search for suitable science and technology achievers: Donald K. Crowdis, from Nova Scotia, and Boyde Beck, from Prince Edward Island. Their assistance was invaluable. Together, they prepared nominations, including biographical information and a list of resources.

To select the achievers who would appear in the exhibition and the book from this long list of nominees, a committee consisting of professional scientists and historians, most of whom were also members of Science East, was established; I had the honour of chairing the committee. The committee decided to include not only those who had been born in the Maritimes but also those who had lived and worked extensively in the Maritimes during their careers and Maritimers who had made their reputations elsewhere.

Discovering these remarkable innovators has fascinated all those who have been associated with the Maritime Achievers undertaking. The human stories and the remarkable accomplishments of these scientists and technologists in our region illuminate a vital aspect of our historic legacy, engage us in understanding the very human process of scientific discovery, and display the remarkable ingenuity for which Maritimers are famous.

Science East sees the touring exhibition as but a modest beginning. Undoubtedly, the list of Maritime Achievers will grow over time. We encourage you to suggest others, and we hope that the number from all three provinces will grow in the future as additional nominations are received. Please visit our website at www.scienceeast.nb.ca.

Allan Sharp, Professor of Physics and Dean of Science,
University of New Brunswick, Fredericton
Chair, Great Maritime Achievers Selection Committee

GREAT MARITIME ACHIEVERS
IN SCIENCE AND TECHNOLOGY

Dr. Abraham Gesner, c. 1870. NBM

ABRAHAM GESNER

Mention the name Gesner in the Maritimes and the response is likely to be, "Oh yes, isn't he the man who invented kerosene?" True as that is, his accomplishments far exceed this single achievement. Abraham Gesner (1797-1864), born in Cornwallis, Nova Scotia, is identified with the Maritimes by birth and by accomplishments. A geologist, museum founder, inventor, educator, business promoter, pioneer in government initiatives, and medical doctor, as well as a musician and a dreamer, this remarkable man developed a process to produce kerosene and lubricant oils from coal, albertite, and crude oil. He is credited with founding the petroleum refining industry in North America and, therefore, with everything that industry has led to, from oil-lubricated and fuelled machinery to jet-powered airplanes.

Gesner's family, Loyalists who had lost their property in the United States through confiscation, settled in Nova Scotia in 1785. There, on the Bay of Fundy, young Abraham became fascinated by the minerals nearby, gathering and studying them. He began the collection that he later deposited in his pioneering museum in Saint John, which is now part of the New Brunswick Museum.

Encouraged and funded by his future father-in-law, Gesner went to London for medical studies; there he also developed his keen interest in earth sciences, studying at the Royal Institution with Michael Faraday. Returning home to marry, he settled in Parrsboro, where he combined his medical practice with collecting and studying minerals. He taught himself the science of geology and explored Nova Scotia, and in 1836, he published *Remarks on the Geology and Mineralogy of Nova Scotia.*

Geology, initially a side interest, became his chief occupation and love when he was appointed New Brunswick's first provincial geologist in 1838. Gesner travelled widely, and in four years he had located most of the mineral deposits in New Brunswick. As a result of other surveys, he described and mapped rock formations in Nova Scotia and Prince Edward Island.

Turning his attention to hydrocarbons and lamp oil, Dr. Gesner researched a substance called albertite, which is a natural pitch-like bituminous material found in New Brunswick's Albert County. From it, he distilled an oil that burned brightly and gave off little smell or smoke. When he lost the right to use albertite in a lawsuit that hinged on definitions, he went on to discover a method of producing lamp oil distilled from coal. This he called *kerosene.* In 1846, while conducting Prince Edward Island's first geological survey, he demonstrated the substance publicly in Charlottetown.

For centuries, whales had been exploited as an important source of fuel for lamps; in 1846, there were 919 whalers killing countless whales for the production of ten to fifteen million gallons of oil a year. Did Gesner's invention save the whales from

Pressed glass kerosene lamp, 1860s. KLHS

extinction? Perhaps not, but because producing kerosene proved to be easy and inexpensive, it certainly slowed down the slaughter.

Plans for his new oil took Gesner to New York, where development money and market demands promised prosperity. He formed the Kerosene Gas Light Company, obtained three patents, and, with others, opened a factory in 1854. A kerosene boom was underway, and the discovery of crude oil in Ontario and then in Pennsylvania provided an even cheaper source of kerosene after Gesner developed a successful distilling procedure.

A Maritimer who enjoyed telling stories and playing tunes on violin or flute, Gesner was outgunned in New York by fiercely competitive newcomers, including John D. Rockefeller. Disillusioned, he sold his patents, moved back to Nova Scotia, and took up residence in Halifax. Returning to old interests of research and study led to his appointment as professor of natural history at Dalhousie University in 1863. However, he died soon after at the age of sixty-seven, leaving behind his family of seven sons and three daughters. His book, *A Practical Treatise on Coal, Petroleum and Other Distilled Oils,* published in 1861, became a standard reference in its field. It explains the development of natural hydrocarbons, describes the different materials that can be distilled, and tells how to build and operate a refinery. Translation into other languages made the second edition the international authority on the subject, but it was published posthumously in 1865, too late for the author to benefit financially from its popularity.

Abraham Gesner was a pioneer in geological surveying, in the extraction of fuels from mineral sources, and in establishing museums. His work was recognized in his own time, and he influenced other scientists, including William Logan, the founding director of the Geological Survey of Canada; Charles Lyell, the foremost English geologist for nearly half of the nineteenth century; and John William Dawson, who carried on the studies Gesner had begun in Nova Scotia. Gesner received honours as well as memberships in learned societies. Although he did not succeed financially and at times seemed beset by misfortune, his enthusiasm for new projects remained undaunted. The Maritimes have claimed him as one of their own, and he is well known in Canada, having been named a National Historic Person of Canada in 1954. In the field of petroleum science and engineering, however, Abraham Gesner belongs to the world.

ROBERT FOULIS

Robert Foulis (1796-1866) was an engineer, foundryman, scientist, and inventor who also studied medicine, engraving, and the fine arts. Like many others who were born in Glasgow at that time, Foulis left Scotland to find a better life and greater opportunity in the New World, and he settled in Saint John in 1822. There he established the province's first iron foundry, installing machinery in early St. John River steamboats and the first Saint John harbour ferry. However, he is recognized internationally for his invention of the world's first steam-operated foghorn, which proved to be a crucial step in the development of navigational aids.

And thereby hangs a sad tale for Foulis. In 1850, he became interested in using coal to manufacture gas. He bought five coal-mining leases in Albert County, New Brunswick, and two years later, he patented his "Illuminating Gas Apparatus." He used this invention to convert the Partridge Island lighthouse at the mouth of Saint John harbour to gas. In the process of building a gas-manufacturing plant on the island and effecting the conversion, he conceived the idea of a steam-powered fog alarm.

In 1852, he asked permission to test this fog alarm; he wrote, "The Alarm Steam Whistle may, by having a very simple machine attached having certain measured intervals of time between each sound, this would distinguish the locality." Not only would the steam foghorn be much, much louder than the large bells that were in use then, but it would also blast out an automatically

Robert Foulis. HRSJ

Original Fog Horn, Partridge Island, N.B., c. 1880, by Frederick H.C. Miles. JCW/NBM

coded pattern of short and long blasts. This code would warn sailors and at the same time tell them exactly where they were. Foulis received no reply to his request, and while he was waiting, someone stole his plans and had a steam fog alarm — the world's first — installed on Partridge Island.

The new warning whistle worked admirably. It won wide praise, but not for the poor inventor. Finally, an investigative committee of the province's Legislative Assembly reported to the Legislature: "We have therefore no hesitation in stating that, after a careful perusal of the papers . . . that Robert Foulis . . . is the true inventor of the practical application of the Steam Horn or Whistle now in such successful and beneficial operation at Partridge Island." Unfortunately, this recognition came too late and brought him no financial reward. By then, having recognized its worth, an American had patented the steam fog whistle.

Robert Foulis also invented an electric dynamo, a machine to harness tidal power, and a tide machine for sawmill operation, and he designed a beacon light as well as an "amphocratic steam engine." Early in his career in Saint John, he established a School of Arts and helped to found the Mechanics' Institute. He also had a special interest in chemistry, and many of his public lectures highlighted "mechanical and experimental philosophy."

Foulis seems to have had a restless mind, and he did not always follow through on his brilliant ideas. As well, he was hounded by bad luck, fires, and family tragedies. His widely varied business ventures and inventions brought him so little financial gain that, plagued by debts all his life, he died a pauper. Even though his name is all but forgotten outside the Maritimes, he worked at the cutting edge of applied science, and his steam-powered, coded foghorn had a powerful effect on modern shipping.

Titus Smith. NSARM

TITUS SMITH

Titus Smith (1768-1850) was born in Massachusetts and came to Halifax in 1783 with his Loyalist family. A naturalist, surveyor, and agriculturist, he was more than a hundred years ahead of his contemporaries in his view that "progress" made by means of unbridled industrialization is just a "scheme to create great fortunes while rendering the lives of the operatives unenviable." His was an early voice warning that modern man's chief ambitions will destroy humanity's own prospects of survival. The son of a clergyman, he viewed the world as fitted by Divine Providence to support life, and he was appalled at people's drastic interference with it.

Titus Smith was reading easily at four years of age, proficient in Latin and good in Greek at twelve, and working on German and French. He was also well schooled in botany and the conservation of plant and animal life. Trained as a land surveyor, Smith acted as an overseer of provincial roads. He also undertook a one-man expedition to survey the interior of Nova Scotia extensively, searching for arable land and other resources, especially timber. His report covered forests, rivers, geological features, and wildlife, and it was accompanied by ink drawings of plants and an overall map of Nova Scotia that was used for thirty years. Through his studies and travels, Titus Smith probably knew his province better than anyone else then alive.

Smith worked diligently for half a century, doing fieldwork, writing for journals and newspapers, and giving lectures to the Mechanics' Institute. He became well known as a specialist in agri-

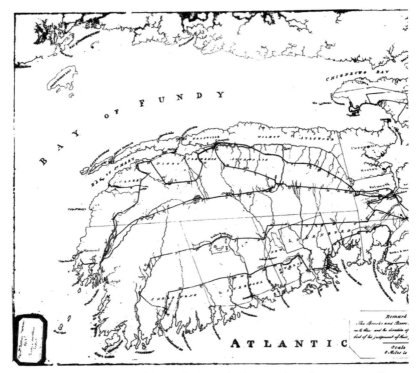

"A Map of Titus Smith's Track through the Interior of Nova Scotia."

culture, a botanist, and a natural historian, always promoting the responsible use of natural resources. Among his most popular publications was the text of *Wild Flowers of Nova Scotia*, a book illustrated by Maria Frances Ann Morris, a Halifax artist.

Smith's advice was sought on many different subjects, and Joseph Howe's newspaper, the *Novascotian*, called him the "Rural Philosopher of the Dutch Village." In an 1835 lecture, Smith spoke of the forests as "the garden of God," where nothing was superfluous or out of place. Smith placed God and nature together; the first concept expressed the personality of God, and the second the force of creation and preservation in the world. Life, he said,

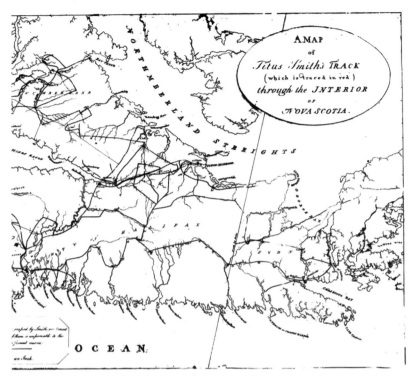

The heavy black lines show Smith's routes through Nova Scotia. NSARM

should accommodate itself to natural rhythms: "Whenever man neglects the dictates of nature, he is sure to be the sufferer."

In 1955, a prominent ecologist characterized Smith's research as "far in advance of his time" and stated that his journals "may well be the first major contribution to ecology in North America." Today's alarm at the systematic degradation of the atmosphere, reduction of the ground water, overfishing of the oceans, destruction of the rain forests, and reckless burning of coal and petroleum would not surprise Titus Smith because it echoes his own warnings regarding careful use of the earth's resources. Today he would be — as he was then — a crusader for Gaia, the living planet.

Moses Henry Perley, c. 1860. NBM

MOSES HENRY PERLEY

Moses Henry Perley (1804-1862) was born in Maugerville, New Brunswick, and grew up in Saint John. He became a lawyer, but from an early age he showed a great enthusiasm for natural history, and he developed into a remarkable self-trained naturalist. An enthusiastic angler, he canoed New Brunswick's rivers and streams, fishing and accumulating knowledge and materials about the physical makeup of the province. Childhood experience and acquaintance with guides led him to value and respect the Native people, whose homes he often visited. These resources, along with his legal understanding, were the foundation of his professional life, and he was recognized as the best authority about New Brunswick's waterways, its inland fisheries, and its other natural resources.

Around 1841, Perley became New Brunswick's commissioner of Indian affairs. His first report, prepared in 1842, backed the Natives in their struggle to oust squatters who, more and more, encroached on their reserve lands; Perley recommended that the Crown continue to hold these lands in trust for the Natives against the interlopers. Perley also became the province's emigrant agent in 1843. In that role, he encouraged both tourists and immigrants to come to the province; among his duties was the placement of the many shiploads of destitute immigrants who arrived. Always optimistic, he viewed his position as "an excellent field for promoting the settlement of the country and forwarding its best interests."

By the time of British North America's reciprocity negotiations with the United States in 1854, Perley had become the most widely recognized authority on the fish of the region and one of the most authoritative ichthyologists in North America. His advice proved vitally important in the trade talks, and because of his expertise, he became a fishery commissioner to enforce the terms of the Reciprocity Treaty itself. In 1862, in his role as fishery inspector, he set off on a trip to the Labrador coast aboard HMS *Desperate*. Unfortunately, he fell ill, and, rather than leave his work unfinished, he decided that the captain should not take him ashore. He died on board the ship, and his body was buried at Forteau, Labrador.

Perley contributed immeasurably to the intellectual life of his community, his province, and his country. He wrote on many subjects, including Native life, past and present; the freshwater fish of New Brunswick and Nova Scotia and the ocean fishery of the Gulf of Saint Lawrence; New Brunswick as a destination for visitors and immigrant; and the exploitable natural resources of the province, including forests and minerals. In 1838, he was a leading figure in forming the Saint John Mechanics' Institute, where he often lectured on natural history, and he was a founder of the Natural History Society of New Brunswick.

ANDREW DOWNS

Andrew Downs (1811-1892) established the first zoo in North America north of Mexico. Born in New Jersey to a Scottish family who moved to Halifax in 1825, he was already a committed student of natural history when he came under the influence of John James Audubon, who visited Nova Scotia in 1833. Taxidermy provided Downs's livelihood, and from 1851 to 1867, it brought him fame and prizes at international exhibitions. His lifelong interest, however, was in keeping live animals for public display, and he is best remembered for his extensive zoo in Armdale, at the head of the North West Arm in Halifax, which he established in 1847. The next public zoological garden to open in North America was in Central Park, in New York City, sixteen years later. At its most extensive, Downs's zoo covered 100 acres and included a greenhouse, an aviary, and an aquarium. By 1860, it was a popular destination for visitors and Halifax residents alike.

Major General Campbell Hardy, an ardent sportsman stationed with the British army in Halifax, visited Downs at his zoological gardens. Downs told him, "There are days when the light seems to bring out the colors on birds' feathers which you would never see in dull weather; days when all Nature seems brightened up by the peculiar state of the atmosphere; when the trees are greener and when the sky has a greater softness and depth than commonly, and your own feelings are in tune all around. . . . Look at that wild turkey as he comes swelling along! The sun's rays light up the

Andrew Downs. NSARM

Andrew Downs in front of his greenhouse and aviary. NSARM

wonderful metallic hues on his neck, back and sides, hues of bronze and green and orange copper, which now and then flash with the brilliancy of the hummingbird's plumage." Then Downs called Hardy's attention to "a pair of pigeons with delicate plum-bloom color on their necks and breasts, a moment later burning with emerald green as they turn to catch a new light, and in another the sparkling tints of hyacinth or topaz."

Downs served on the governing committee of the Halifax Mechanics' Institute and as curator of the institute's collection, and he was intent on establishing a provincial museum of natural history. To help secure the necessary government support, Downs became curator of the institute, and he established his zoological gardens at least in part as a study collection for such a museum. He was a founder of the Nova Scotia Institute of Natural Science in 1862, and he wrote and read papers before the society. He delivered an updated version of one of these, his list of Nova Scotia birds, before the Royal Society of Canada in 1888. This is the best known nineteenth-century study on the subject.

Although Downs's chief interest was in wild animals, he investigated breeding domestic animals to increase their economic potential. He imported poultry from England to upgrade his own breeding stock and took prizes in provincial competitions. Encouraged by this success, he adapted English pheasants to the Nova Scotia climate, and he also imported Chinese sheep.

The fame of Downs's zoological gardens spread, and zoologists at the Smithsonian Institution became aware of his achievements. In 1868, at their suggestion, he was invited to assume administration of zoological operations at Central Park in New York City. Excited at the prospect of continuing his work on a larger scale, he auctioned off his Halifax collection and property and moved to New York. However, the promised position did not materialize; American naturalist Charles Hallock diagnosed the problem as "political jugglery." Downs returned to Halifax, bought property near his former site, and tried to re-establish his zoological gardens. This time he did not succeed, and he sold his new venture in 1872.

A member of the Zoological Society of London, Andrew Downs personified the mid-nineteenth-century interest in science, public parks, and recreation. In 1864, he was described as "one whose heart is in his work, if ever a man's was, and who has the liberality of spirit possessed by all true lovers of nature." In his tribute, Campbell Hardy said, "Every visitor desirous of acquaintance with wild life in the woods or by waters of Acadie went to Downs for advice or reference."

BENJAMIN TIBBETS

Benjamin Franklin Tibbets (1813-1853) invented the world's first compound marine engine. Born in the tiny settlement of Mill Cove, in Queens County, New Brunswick, and raised in humble circumstances, Tibbets received only four years of schooling, followed by technical training as a watchmaker. With his ever inquiring mind and driven by an indomitable perseverance, this remarkable New Brunswicker became a master watchmaker, a professional painter, and an accomplished musician, as well as a marine architect and inventor. Tibbets devised a new system for combining high and low steam pressures into a unique type of compound marine engine, the first of its kind.

Early in his youth, Tibbets became known for his wonderful knowledge of the mechanical arts, but steam power held a special fascination for him. He began by making miniature engines that actually worked. His contemporaries recalled that he talked often of steam-powered machinery, and they told the story of a small engine Tibbets had created, when still a teenager, to demonstrate steam-generated power. He continued to study the steam engine as an apprentice in workshops and foundries in Canada and the United States, and there he also developed his amazing mechanical talents.

Keeping enough fuel on board was a constant problem for steam-powered ships, and so Tibbets turned his attention to marine engines. He devised a way to use the steam twice, first under high

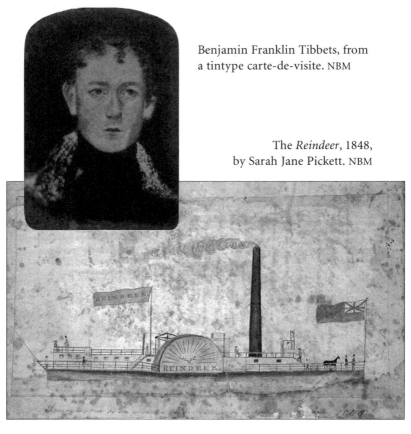

Benjamin Franklin Tibbets, from a tintype carte-de-visite. NBM

The *Reindeer*, 1848, by Sarah Jane Pickett. NBM

pressure in one cylinder, and then a second time, in another cylinder, when the pressure had decreased. The first engine of his design was installed in the St. John River steamboat *Reindeer*. Tibbets's genius and inventiveness extended to supervising the building of the engine and to designing and overseeing the construction of the steamboat's hull. Thus he created a boat that was superior to previous steamers on the St. John. Some said that he had studied the Maliseet birchbark canoe and incorporated its buoyancy principles into the design of a vessel that would offer the least possible resistance to the water. That may or may not be true, but, as a contemporary newspaper reported, "The *Reindeer* is neatly built, handsomely painted, and moves through the water swiftly and with swan-like gracefulness." The engine was highly successful, establishing records for speed, fuel economy, and endurance, and that same engine was used for more than seventy years, in three different vessels, outliving all of them.

Other engines of Tibbets's design followed, both in New Brunswick and in Montreal, where his uncle owned a small machine shop capable of manufacturing them. Tibbets successfully obtained patents for his inventions in both New Brunswick and Lower Canada, as Quebec was then called. As well, in 1846, the New Brunswick Legislative Assembly voted him £100 "in consideration of the great talent and ingenuity exhibited by him" in the hope that he would apply for a patent in the United Kingdom.

Unfortunately, Tibbets was plagued by a lack of funds and ill health; pulmonary tuberculosis forced him back to New Brunswick, where he died in his boyhood home at the age of forty-five. Canada's Historic Sites and Monuments Board declared Tibbets and his invention of the first compound marine engine to be of national significance. J.J. Brown published a history of Canadian invention in which he calls Tibbets "a genuine marine pioneer, of whom all Canadians may well be proud." Eighty years after his death, the local historical society erected a monument in his honour.

Charles Fenerty. NSARM

CHARLES FENERTY

When Charles Fenerty (1821–1892) invented the first process for making paper from wood pulp in 1838 or 1839, it was the sort of development sometimes imagined as "inventing itself." Paper was not new; it was well known as a flexible, durable surface on which records could be written or drawn — not carved, as on wood, stone or clay. And paper made out of rebonded fibres from plant sources was not new either. The Chinese, Egyptians, and other ancient civilizations had kept records on paper made from papyrus, cotton, linen, and hemp. Of course, paper nests made by wasps and hornets were taken for granted, but they held all the clues necessary for making paper from a source as bountiful as wood.

Fenerty's family operated three sawmills on the Windsor Road, outside Halifax, and as a youth, he was fascinated by the process, especially by the waste sawdust and fine powder from the saws. When he was less than twenty years old, Charles visited the Holland paper mill, near Bedford Basin. Apparently that was when he learned that reworked rags, the accepted source of papermaking material, were showing signs of becoming scarce; it was difficult for the mill to obtain a regular supply.

The naturalist Titus Smith, a family friend, had written that spruce trees could probably be useful for making paper. Taking the hint, Charles Fenerty started to experiment. Choosing soft spruce as the most suitable, he produced what he called "pulp paper" from sawmill residue. Tradition has it that he realized that the lumber mill waste could become the primary resource in making paper

The Saint John Pulp and Paper Company, Mispec, New Brunswick, c. 1910. Charles Fenerty lived to see mills like this using his concept for making paper from wood pulp. NBM

and that young Charles, at seventeen or eighteen, showed a paper sample made from wood fibre to his future brother-in-law, Charles Hamilton. If this is indeed true, Charles Fenerty clearly is the inventor of wood-pulp paper.

The realization could have made him a discoverer of world significance, but it wasn't until several years later, on October 26, 1844, that he wrote a letter to the *Acadian Recorder* on a sample of the paper he had made. It was, he said, "as firm in its texture, as white, and to all appearances as durable as the common wrapping paper made from hemp, cotton, or common materials of manufacture, [it] is actually composed of spruce wood reduced to pulp and subjected to the same treatment as paper."

Fenerty might have tried to benefit financially from his invention. His family's successful mills demonstrated their business acumen, and they could have provided the raw material, adapting their machinery to the process. However, his letter to the *Acadian Recorder* ends as he leaves his papermaking experiment "to be prosecuted by the scientific or the curious." He was certainly curious himself, and his interests ranged widely; it is not known why he did not attempt to seek a patent. Once he had made the discovery and proven its efficacy, his interest moved on to other things.

Charles Fenerty — a handsome fellow, it is said — fell in love with Charles Hamilton's sister and wished to marry her. When she rebuffed him, Fenerty went off to Australia, possibly to the gold rush, later working in Melbourne as a journalist. When he heard that the engagement of another Hamilton sister and his own brother was broken off, he rushed back, proposed to the girl, and married into the Hamilton family anyway. He resumed working in his own family's lumbering business, served locally as health warden and overseer of the poor, and returned to an abiding interest in poetry. He wrote on such varied subjects as a giant tree on the family farm, on Confederation a year before it happened, on the decaying Prince's Lodge, on Polish and Italian independence struggles, and on the spread of British culture around the world. His poems were published, and some won prizes.

In the meantime, while Fenerty did nothing with his discovery, a German weaver, Frederich Keller, invented the same papermaking process. Fenerty saw this process in use for twenty-five years before his death, at which time two pulp paper mills were operating in Nova Scotia. Fenerty received no credit for his previous invention, and neither he nor anyone else in the province had taken commercial advantage of it; in fact, no one even sought a patent.

Only in 1926 did the Nova Scotia Historical Society recognize Charles Fenerty's invention of a process for making paper from wood, and they placed a plaque at his birthplace. In 1987, Canada Post issued four stamps to honour scientists and technologists, Fenerty among them as the pioneer behind the pulp and paper industry.

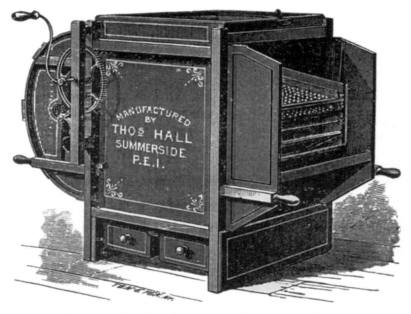

Eureka Fan Mill and Seed Separator. There is no implement on the farm better than a good fanning mill. The most careful preparation of the soil will not bring the desired result unless the seed has also been carefully prepared before sowing. The Eureka is a combined Fanning Mill and Seed Separator; that is it will take either grain or grass seed from chaff and prepare it for either market, seed or exhibition. It has never been beaten in any trial, and we do not know of any mill sold that can equal it. Give it a trial and you will be satisfied.

— Hall's *Descriptive Catalogue*, 1893. PEIMHF

THOMAS HALL

The isolated farms of yesteryear encouraged ingenuity as they enforced hard work. Any invention that made work easier or more productive was welcome, and good farm mechanics were common and appreciated. They could sharpen scythes, make tools, pound hot iron into shape, build furniture, restore old nails and hinges, and generally make things work and last. Many a famous inventor and manufacturer began as a farm boy, including Thomas Hall (1836-1919) of Wilmot, near Summerside, Prince Edward Island.

Hall took to the agricultural implements about him very early and, self-taught, soon not only kept farm machines in shape but improved them. By the age of twenty-four he had established a small but viable shop in Summerside. He could repair anything, and soon he ventured into the design and manufacture of new farm equipment. At first, he tended to produce improved versions of machinery made by established companies, always aiming for simplicity, practicality, and durability. Urged on by losses in a fire in 1873, he built a better shop with steam-powered manufacturing machinery. Soon, he began to show his products at agricultural fairs not only on the Island but on the mainland. The Hall Manufacturing Company became a popular farm implement manufacturer, particularly renowned for its threshing machines.

Hall's expanded line came to include a horse treadmill with a level bed, a combined thresher and cleaner, a ratchet lever jack, the

Thomas Hall. PEIMHF

Level Tread Horse Power.
The level-tread planks, or lags, of this machine are made
of rock maple and are 2 ½ inches thick, dressed on both sides;
they are 7 ½ inches wide, and the corners are taken off to prevent
the horses from catching their calks in the joints.
They are connected by iron links at each end.

— Hall's *Descriptive Catalogue*, 1893. PEIMHF

Eureka Fan Mill and Seed Separator, and other implements, all promoted in an illustrated catalogue. These machines sold well, thanks to the reputation they enjoyed, and Hall received many glowing testimonials from well-pleased customers. The threshing machine was particularly successful, winning first prize at the 1881 Dominion Exhibition in Halifax. The Summerside *Pioneer* remarked, "Mr Hall, though so quiet and unassuming in appearance and manner, occupies now the proud position of standing at the head of all mechanics in his line throughout the Dominion."

At the time of Confederation, Hall had wanted to expand his business, but he was so far from the large Canadian markets to the west that he could not do so. Indeed, he was not the only one in the Maritimes to discover that his own market faced competition from outside, especially from Ontario, where protective tariffs supported manufacturers. As well, Hall himself had found horse-powered machinery so satisfactory for farm work that he continued to specialize in it. Even when he retired in 1914, his company remained strictly a manufacturer of horse-powered implements. The Hall Manufacturing Company continued to make much the same machinery into the 1950s, and the famous thresher designed by Hall had a lifespan of some sixty years.

Francis Peabody Sharp in his orchard with a dwarf apple tree, 1875-1900. NBM/JAD

FRANCIS PEABODY SHARP

Francis Peabody Sharp (1823-1903), the apple king of New Brunswick, excelled in many fields: he was an orchardist, a pomologist, an inventor, a businessman, and a musician. Canada's first apple breeder, he had a scientific mind, and though largely self-trained, he became the foremost authority on marketable fruits as well as a successful businessman in a very competitive field.

Born on a farm near Woodstock to a prosperous family, he received early home-schooling from his gifted mother before he went to grammar school. His well-read parents encouraged young Frank to study on his own, and keen curiosity marked his teen years. Science captivated him. He discovered the joy of understanding the insights revealed by science, and this joy influenced his way of observing and thinking.

Frank Sharp had a passionate interest in fruit trees, and by the time he turned twenty-three, he had established a prosperous business, Woodstock Nurseries. There, he conducted patient and careful research into developing varieties of fruit such as apples and plums that would do well in the climate of central New Brunswick, which had generally been thought too cold for fruit of good quality.

Through striking foresight and tenacity, Sharp became a superb experimentalist. His specialty was apple culture. "A few persons, it is true, had a few trees, raised here and there for their own use, but there was not an apple raised for market until I did it," he observed, reminiscing about his early years as an orchardist. "I learned it *de novo*; I started at first principles. At first I knew nothing, even about grafting. A man from Saint John told me about grafting. Father had quite an orchard of seedlings planted across the river. I bought out the property and grafted the trees all over with good varieties." Sharp crossbred from trees as far away as Russia in his quest for cold-hardy fruit. He applied his studies, experiments, and correspondence with experts in Canada, the United States, and Great Britain to tree grafting, growing trees from seed, and especially to the production of new apple strains by plant hybridization.

In 1854, Sharp introduced the New Brunswick apple, popularly called the New Brunswicker, which took first prize in competitions with apples from all of the eastern United States. It was followed by the Fameuse, the Crimson Beauty, and others that attained widespread popularity. The Early Scarlet, the earliest red apple at that time, has since been used in breeding programs the world over. In his trials, Sharp prepared lists identifying which varieties would do best where. "Even in so small a country as New Brunswick," he said, "as much as three separate lists will be required. Some fruits do well at Fredericton, which here in Woodstock are of no value whatever, and at Sussex the St. Lawrence apple appears to be at home while here it is of no value. So it is not to be wondered at that Nova Scotia varieties and those adapted to New England should be totally useless here."

His experimentation with different varieties was accompanied by equal attention to orchards. Sharp planted trees close together, fertilized them heavily, and selected and pruned the trees to keep them small as he grafted his imports. The result was heavier yields of fruit that was larger and of better quality. By 1878, he had 10,000 trees growing in orchards, 300,000 more in nurseries, and an additional 200,000 in storage, 150,000 of which were already sold. He shipped apples, cider, and cider vinegar by the carload to the United States and to Canadian destinations as far away as Manitoba. By this time, Sharp's agribusiness included plums, rhubarb, and vegetables such as cabbage; all were new varieties produced by his experimentation. Superb manager that he was, he branched into maple syrup, barrel making, canning, and packing.

Sharp achieved recognition in his own time, and the renowned Brogdale Horticultural Trust of England designated him "Canada's first fruit breeder." However, he declined the many offers this recognition attracted, including a faculty position at Iowa State College and an opportunity for a knighthood. The Government of Canada commissioned him to write a book on fruit culture, but he lost the manuscript and his volumes of research notes in a fire. Two scholarly papers that he read before groups in Ottawa and Fredericton have survived, and these provide some insights into his achievements.

After sixty years of prosperity, a series of personal tragedies, including a major fire and the deaths of three of his children in one week, brought an end to his remarkable career. Without his leadership, growers tended to drift back to importing apple varieties proven successful elsewhere, and many of his new developments became obsolete. Even so, through his industry, singleness of purpose, and practical experimental techniques, Sharp transformed apple growing in the region from a hobby to an industry.

At the time of Sharp's death, Tappan Adney, his son-in-law, observed of his career: "Such was the strong scientific bent of his mind that he expended his means in experiments designed to prove of commercial value to the world and gave results of his knowledge freely in the form of addresses and of contributions to the press. He was regarded as one of the most eminent investigators into plant life as applied to fruit growing in northern latitudes and made a fruit-growing country of a province not deemed to be capable of raising fine fruits of any kind." Of the man himself,

Adney said, "His mind seemed ever bent to fathom nature, to discover the utmost possibilities of good to men lurking in earth and air, in soil, stalk, root, leaf, and blossom."

King's Landing, the New Brunswick historic settlement, has begun to establish an agricultural memorial to Francis Peabody Sharp. There, his creations will be grown in experimental orchards, and a replica of his nursery is being built. This will be a fitting memorial to the man who dominated the commercial orchard and nursery business in Canada and who influenced apple growing throughout North America.

FREDERICK NEWTON GISBORNE

Frederick Newton Gisborne (1824-1892), a telegraph engineer, invented a system of submarine cable insulation that made underwater telegraphy feasible. Gisborne was born in England and, while still a teenager, he successfully experimented with natural rubber, in the form of gutta percha, to obtain an electric wire coating resistant to water. In 1845, he immigrated to Canada. At first, he worked for the Nova Scotia Telegraph Company as a telegraph operator, but he rapidly rose to the position of manager.

By 1847, Gisborne was general manager of the British North America Electric Telegraph Association, a company formed to build a telegraph line between the Maritimes and Upper and Lower Canada. He directed this project to its successful conclusion, and he also oversaw building connections with the United States. Still dreaming of a submarine telegraph cable, however, he established a company in 1850 to demonstrate that his technology would work. Employing the insulated wire he had designed, his company laid a cable on the bottom of the Northumberland Strait between Cape Traverse, New Brunswick, and Cape Tormentine, Prince Edward Island, a complex task completed in November 1852. If it had been in place six months earlier, it would have been the world's first underwater telegraph cable; even so, it was the first in North America.

The New Brunswick-Prince Edward Island cable was Gisborne's first step toward the achievement of his new dream: a trans-

Frederick Newton Gisborne. VUL

Atlantic cable. Next, he attempted to lay cable across the Cabot Strait between Nova Scotia and Newfoundland and across Newfoundland to St. John's. This attempt was fraught with problems, including a sojourn in debtor's prison, but in 1856 the project finally succeeded. Meanwhile, Gisborne joined the Atlantic Telegraph Company to promote and participate in the laying of a cable under the Atlantic ocean. Between 1858 and 1866, three cables broke, but, thanks to the technology Gisborne designed, the first telegraph message passed between Europe and North America in 1859, and the continents were finally linked by telegraph in 1866.

Gisborne's Newfoundland difficulties turned him away from telegraphy for a while and towards mining. He became the president of the Mining Association of Newfoundland, and he went to England as mines and minerals agent for the Nova Scotia government. Despite this diversion, however, telegraphy was always Gisborne's main interest. In 1879, he became superintendent of the Dominion government's Telegraph and Signal Service, and in that position he worked on a telegraph cable network intended to span the Pacific to Australia by way of Canada's northwest, the Aleutian Islands, Japan, and New Guinea.

Frederick Gisborne was a charter member of the Royal Society of Canada and a member of the Council of the Canadian Society of Civil Engineers and other scientific associations. His pioneering inventions for insulating underwater telegraph cable and his efforts to create a worldwide telegraphy system make Gisborne a landmark figure in science and technology.

Simon Newcomb. NSARM

SIMON NEWCOMB

Simon Newcomb (1835-1909), born at Wallace Bridge, Nova Scotia, became a world-renowned astronomer and mathematician. His advances in the field of celestial mechanics were fundamental to navigation in his own time and to the scientific research of the future.

The wanderings of Newcomb's family were typical of the times. In the nineteenth and well into the twentieth century, hardly a Maritime family was without connections in the "Boston States." Many came from there in times of trouble, and many went there in search of prosperity. Simon's parents fit into both categories: they came from New England as teachers, but, not faring well in Nova Scotia or New Brunswick, they returned to the United States.

Newcomb spent an unhappy, poverty-stricken childhood in the Maritimes. He showed great talent in mathematics and a strong interest in astronomy, but after a mental breakdown when he was only seven years old, he stopped attending school. He went to work as a farm labourer to strengthen his body and improve his manual skills, but he hated that occupation. As a teenager, he was apprenticed to a Moncton herbalist, but when he realized that this man was a fraud, he ran away to Calais, Maine. By this time, his widowed father lived in Salem, Massachusetts, and Simon made his way there by working for his passage aboard a ship.

Largely self-educated, Newcomb taught school in Maryland in 1854 and 1855. Always keenly interested in mathematics and science,

Participants in an expedition to observe the 1874 transit of Venus, who had gathered to practice on the grounds of the United States Naval Observatory. Simon Newcomb is seated in front. USNOL

he corresponded with established scientists, particularly Joseph Henry, who was secretary of the Smithsonian Institution and the most important physicist in the United States. He showed such promise that doors began to open to him. Henry found him a position as a "computer" with the *American Ephemeris and Nautical Almanac*, a branch of the navy, in Cambridge, Massachusetts. While he worked for the *Nautical Almanac*, he enrolled at Harvard. He excelled in his studies, published his first important research, and graduated summa cum laude in 1858.

In 1861, Newcomb became professor of mathematics at the Naval Observatory in Washington, D.C., and he also held a professorship at Johns Hopkins University in Baltimore, Maryland. His research, imagination, and careful supervision of his human "calculators" enabled him to revise and standardize the methods for establishing lunar and planetary positions and thus bring order to the *Nautical Almanac* tables. This had never been done before, and it was his greatest contribution to science.

This former Nova Scotia child labourer kept in touch with his cousins in Canada while he accumulated national and international honours. During his lifetime, he was recognized as the greatest American astronomer. The author of more than five hundred publications, he became president of the American Association for the Advancement of Science. Among his many awards were the Dutch Huygens Medal, the Bruce Medal, awarded by the Astronomical Society of the Pacific, and the Copley Medal of the Royal Society of London. The Naval Observatory conferred upon him the rank of rear admiral (retired), and numerous honorary doctorates and decorations were heaped upon him in North America and Europe. Albert Einstein benefited from Newcomb's great contribution to science, using his work to develop the theory of relativity. In 1935, Newcomb was named a National Historic Person of Canada. Until 1984, his revision of the *Nautical Almanac* tables and his work on planetary orbits remained fundamental to the nautical and astronomical almanacs of the United States and Great Britain.

Sir John William Dawson. MUA

JOHN WILLIAM DAWSON

Pictou, Nova Scotia native John William Dawson (1820-1899), a committed Christian, contemplated entering the Presbyterian ministry, but, taking natural history to be the work of God, he became instead a scientist of worldwide reputation. Dawson distinguished himself in every field he entered: geology, paleontology, and teaching, and writing and publishing on these topics. He was the only person ever to serve as president of both the American and the British Association for the Advancement of Science, and in 1884 his contributions to learning and to society were rewarded with a knighthood.

Always keenly interested in education, Dawson became the first superintendent of education for Nova Scotia in 1850. He established the first normal school in Nova Scotia at Truro and served as its principal and science instructor for nineteen years, teaching chemistry, agriculture, geology, zoology, and botany.

Dawson built an impressive collection of minerals, shells, fossils, and other natural objects, and he exchanged specimens with Abraham Gesner and various naturalists in Boston. He formed a lifelong friendship with the renowned scientist Charles Lyell, with whom he conducted field investigations; they published many of their findings in the *Quarterly Journal* of the Geological Society of London.

In 1855, William Dawson became the fifth principal of McGill College, and over the next thirty-eight years he transformed McGill into one of the continent's leading universities. A modernist in both science and education, he admitted women to McGill, and he established close relations with the Smithsonian Institution in Washington, D.C. and the Museum of Comparative Zoology at Harvard. Princeton University offered to make him a professor of natural history, but Dawson chose to remain in Canada. One reason was that Peter Redpath agreed to endow a major natural history museum if he would stay at McGill. The result is the Peter Redpath Museum, which still dominates the university's lower campus. By the time Dawson retired, the *Times* of London said McGill had won a place "second only to Harvard in North America."

Dawson was involved in several geological controversies, most notably the one that raged around Darwin's theory of natural selection. However, his scientific credentials were so impeccable that, although he took strong issue with *The Origin of Species*, Darwin himself co-signed the testimonial putting forward Dawson's name for election as fellow of the Royal Society of London. In 1882, under the patronage of the governor general, the Marquess of Lorne, Dawson and others formed the Royal Society of Canada, and Dawson became the first president.

Dawson's biographer, D.J.C. Phillipson, writes: "Principal Dawson's career spanned the transformation of science from a fixed curriculum of 'natural philosophy' to an array of professional disciplines focused on research. Like his predecessors, he wrote on subjects from farming to philanthropy, but he also earned a solid reputation as a geologist, equally at home on a cliff, chipping out samples, or in his study, synthesizing and interpreting the processes of geological time. He was the leading expert of his day on early fossil plants and took special pride in his identification of *Eosoön canadense* as a coral, thought to be the oldest non-plant fossil known. Always controversial, it was not for another fifty years that *Eosoön* was shown to be a rare crystal formation rather than a living animal."

John William Dawson (centre front, with white beard) and other geologists in a quarry. MUA

In addition to his knighthood, Sir William received countless honours and awards. His books on geology and natural history include the monumental *Acadian Geology* (1855), a practical and authoritative guide to the geology and economic resources of the Maritime provinces; *Handbook of Canadian Zoology*, 1871; and *Handbook of Canadian Geology*, 1889. The tribute in the 1900 Royal Society *Transactions* says, "He was a constant contributor to the scientific publications of the two continents for more than half a century.... As a writer he was remarkably lucid, able to invest the most scientific subject with deep interest and make it intelligible to the most ordinary reader." He was named a National Historic Person of Canada in 1943.

In his presidential address to the Royal Society of Canada, Dawson described the aims of the society: "We should prove ourselves first unselfish and zealous literary and scientific men, and next Canadians in that widest sense of the word in which we shall desire, at any personal sacrifice, to promote the best interests of our country and this in connection with a pure and elevated literature, and a true, profound, and practical science." John William Dawson himself exemplified these aims.

DAVID HONEYMAN

Although the eighteenth and nineteenth centuries were notable for continual faceoffs between established religion and investigations in the natural sciences, Dr. David Honeyman (1817-1889), a Presbyterian clergyman and geologist, suffered no such difficulty. He believed that the Deity was consistent and that humans should investigate his works as fully as his revelations.

Honeyman was born in Scotland and studied natural science and oriental languages at the University of St. Andrews, but then he chose the church as a profession and devoted his energy to theology. When he moved to Nova Scotia in 1848, service in the ministry was uppermost in his mind. Ten years later, however, Honeyman resigned his charge at Antigonish, Nova Scotia, choosing instead to follow his passion for scientific research and geology. The flames of this passion had been fanned by John William Dawson's landmark book *Acadian Geology*.

From 1861 to 1882, the Nova Scotia government employed Honeyman to prepare geological, botanical, agricultural, and fisheries displays for international exhibitions in London, Dublin, Paris, and Philadelphia. For this work, Honeyman received rave reviews and international recognition. In the mid-1860s, he tried to secure a commission to investigate and record the geology of Nova Scotia through a government-sponsored survey, similar to the one that had been established in the Province of Canada

David Honeyman. NSARM

(present-day Ontario and Quebec) in 1842. Charles Tupper had promised to establish such a survey in Nova Scotia, but, perhaps due to the pressures of approaching Confederation, he was not able to keep his word. Instead, in 1864-1865, the province commissioned Honeyman to prepare an overview of Nova Scotia geology. In 1868, he worked on the Geological Survey of Canada under Sir William Logan. Eventually, he produced more than fifty published articles on natural history, especially geology and marine life.

The establishment of the Provincial Museum of Nova Scotia in 1868 was Honeyman's greatest achievement. The museum held the collections he had gathered for the international exhibitions, as well as those of the by now disbanded Halifax Mechanics' Institute. To enable people to study the museum collection, Honeyman wrote *Giants and Pigmies Geological*; it was published in 1887. He served as the museum's curator until his death in 1889.

Honours were heaped on Honeyman at home and abroad. A founding member of the Geological Society of America and one of the first fellows of the Royal Society of Canada, he was also secretary and editor of the Nova Scotia Institute of Natural Science and a member of the Geological Survey of Canada. Like many nineteenth-century scientists, David Honeyman was an amateur who, largely through personal interest and independent study, became an expert scholar and helped lay the foundations for a new science.

JAMES ROBB

An educator and a scientist, James Robb (1815-1861) created a highly effective teaching museum which lives on in the University of New Brunswick's outstanding Connell Memorial Herbarium. His publication of *Agricultural Progress: An Outline of the Course of Improvement in Agriculture . . . with Special Reference to New Brunswick* resulted in the establishment of a provincial Board of Agriculture, with Robb as its secretary.

Born in Scotland, Robb attended the University of St. Andrews and then studied medicine at the University of Edinburgh. He also spent some time on the continent, where he collected natural history specimens and met with leading scientists in the fields of medicine and botany.

In 1836, he accepted an appointment to King's College (later the University of New Brunswick). He arrived in Fredericton in September 1837, as the college's first professor of chemistry and natural history, a subject that included botany, zoology, physiology, anatomy, geology, and mineralogy. Unfamiliar with New Brunswick, he soon undertook an extended exploration, walking and canoeing through the backwoods collecting minerals, fossils, and other geological specimens. Thus, although he was a very recent immigrant, he educated himself well enough to teach his students about their own natural surroundings.

Robb gained a wide reputation as a botanist and was prominent in the work of the American Association for the Advancement

The University of New Brunswick's Connell Memorial Herbarium, with an ox-eye daisy specimen collected by James Robb in July 1843 in a Fredericton grassland. ROGER SMITH

of Science. His botanical collection, gathered in Europe, vigorously enlarged during his travels in New Brunswick, and supplemented with specimens donated by Abraham Gesner, is a legacy of lasting importance.

In 1849, Robb accompanied J.F.W. Johnston, a British agricultural scientist appointed by the province of New Brunswick, on a two-thousand-mile study tour to survey the state of the province's agriculture. The next year, Johnston submitted his report, *The Agricultural Capabilities of New Brunswick*, which included Robb's geological map. By establishing the true stratigraphic positions of certain rocks in large areas of the province, Robb's map confirmed Sir Charles Lyell's controversial thesis that these formations were in fact older than coal formations. Two years later, Robb helped Abraham Gesner in his unfortunate lawsuit by corroborating Gesner's contention that albertite is not coal but a different mineral entirely.

James Robb. UNB/HIL

Without neglecting his teaching responsibilities, Robb led a movement to establish the New Brunswick Society for the Encouragement of Agriculture, Home Manufactures, and Commerce, believing as he did that agriculture, scientifically pursued, was and would remain fundamental to the economy of New Brunswick.

When Robb died of pneumonia, his death was blamed on the damage his strenuous life had done to his health. Amid the widespread mourning, his friend and colleague, William Brydone Jack, called his passing a public calamity. Had Robb lived, he might well have become the first president of the newly constituted University of New Brunswick.

William Brydone Jack, during his term as president of the University of New Brunswick (1861-1885). UNB/HIL

WILLIAM BRYDONE JACK

William Brydone Jack (1817-1886), British North America's pioneer astronomer, came to teach at King's College in Fredericton, New Brunswick in 1840, when he was only twenty-three years old. His academic preparation in his native Scotland had been stellar; he excelled in mathematics, natural sciences, and philosophy, and he had taken the highest prizes at St. Andrews University. As persuasive and practical as he was brilliant, he focused especially on the application of science to the immediate problems and opportunities he saw before him.

Brydone Jack pushed for emphasis on engineering, mathematics, and natural sciences, but astronomy was his passion. He persuaded the college to allocate the entire 1850 science budget of £450 to the purchase of a state-of-the-art refracting telescope, "an equatorial telescope of 6-inch clear aperture and 7 1/2 feet focal length, made by Mertz & Son of Munich." The next year, he insisted on an additional £200 to build an observatory to house it. A stalwart promoter of practical education, he succeeded in 1854 in getting funding for the first civil engineering chair in British North America.

Brydone Jack is credited with solving the huge problem of inaccurate land surveying that existed in the colony. In 1858, he described this problem in a lecture to the Fredericton Athenaeum: "Hitherto the wild lands of New Brunswick have been parcelled out without reference to any general or comprehensive plan and with little or any unity of system. The consequence has been that

The William Brydone Jack Observatory, University of New Brunswick, 1904.
UNB/HIL

A transit telescope used by William Brydone Jack, now housed in the observatory museum at the University of New Brunswick.
MARIE MacBEATH

titles are often found to conflict with each other, and in some cases separate and distinct grants have, I believe, been made of the same premises. Moreover the evil is increasing every day; and the labour thus entailed on some of the gentlemen belonging to the Crown Land Office, and the trouble and ingenuity required for patching together and reconciling the discordant and imperfect surveys put into their hands can scarcely be imagined." After a decade of work, he had found a solution. In the process, he had devised a means for standardizing surveyors' chains and created a standards laboratory for surveying equipment, the first of its kind in British North America. As well, he was the first to ascertain the true longitudinal coordinates for both Fredericton and Saint John.

When King's College became the University of New Brunswick, William Brydone Jack became in effect its first president, a position he held until he retired in 1885. But this did not change his professorial tasks. Still teaching astronomy, engineering, surveying, and mathematics, as well as giving public lectures, especially in astronomy, he continually pressed for better standards in high-school science teaching. At the same time, he did not neglect classical studies; rather, he introduced the "Scottish system of combining classical studies with a scientific and practical education." Nor did he neglect his university administrative duties. Whether enlarging the faculty with outstanding young scientists, adding to the buildings, or recruiting new money, especially for scientific research, he set the pace as the Maritimes joined the other British North American colonies in Confederation.

Brydone Jack's biographer, J.E. Kennedy, says: "William Brydone Jack's contributions to education in New Brunswick extended over forty-five years and included service on the provincial board of education from 1872 to 1885. To the intellectual life of the university and the province he brought the best traditions of his ancestry and education. On his retirement in 1885, the university senate awarded him a pension and in high tribute to his dedicated service appointed him one of its members. His death within the next year, however, deprived the university of his continuing sound advice." The Brydone Jack Astronomy Club is named in honour of his lasting contributions to the university, province, and nation. His astronomical observatory, declared by the Historic Sites and Monuments Board to be the first in British North America, is now a national historic site, serving as a tiny museum of astronomy on the University of New Brunswick campus.

GEORGE FREDERIC MATTHEW

George Frederic Matthew (1837-1923), a distinguished scholar, became a leader in scientific endeavours, internationally known for his studies in paleontology. He was born to a merchant family in Saint John, New Brunswick; although his formal education was limited to the grammar school in the city, he had a natural bent for scientific investigation. He supplemented his reading with observation, examination, and questioning those who were knowledgeable in the natural sciences.

He joined the Customs Service at sixteen, and eventually he rose to become chief clerk and surveyor. He married early, but even as his family grew, he found time to study the rocks and fossils of New Brunswick, Prince Edward Island, Nova Scotia, Quebec, and Maine. The Geological Survey of Canada employed him from time to time, and he became the service's expert on the Cambrian formations of the region.

Because he had few opportunities for travel, he corresponded extensively with scientists at home and abroad, and he became recognized as "a serious, scholarly authority in his scientific field, destined to occupy a lasting place in the history of North American paleontology." Much of his fossil collection and his scientific correspondence has survived and can be found at the New Brunswick Museum; both the collection and the correspondence show him to have been a pioneer in fossil research, an internationally known paleontologist, a respected archaeologist, and a noteworthy linguist.

Firm in his belief that "It is the business of science to discover, record, and classify facts," George Matthew began publishing at the age of twenty-five. He left more than 350 species records of new plants and animals, and he was the author of more than two hundred scientific papers and monographs, including an impressive series on Cambrian fossils found in the Maritimes. These papers were published in many periodicals, including the *Transactions of the Royal Society of Canada*.

Matthew became widely respected and even revered over the course of his long life, helping and influencing younger scientists. He lectured at the New Brunswick Natural History Society, of which he was a founder, the president, and a life member. Among the distinguished members of that society were Charles Frederick Hartt, William Francis Ganong, and Loring Woart Bailey, all of whom made important contributions to the scientific knowledge of New Brunswick and beyond. With his friend Bailey, Matthew mapped the geology of New Brunswick.

Matthew was actively involved in the acquisition of the Gesner Museum by the Natural History Society. Later, serving as curator, he worked to combine this and other collections as the New Brunswick Museum, to which he contributed many of the fossil specimens on which his own paleontological reputation was built. In his view, the objective of a museum was to educate as well as to accumulate and conserve, an ideal he shared with Bailey, who headed the teaching museum at the University of New Brunswick. The interests of these two men ran parallel and overlapped repeatedly, and their constant, fruitful collaboration developed into a deep, lifelong friendship.

One of George Matthew's greatest accomplishments was the early training of his oldest son, William Diller Matthew, who has been called the father of mammalian paleontology. Father and son were working together when William, then in his teens, discovered the giant trilobite fossil, *Paradoxides regina*.

George Frederic Matthew, c. 1889. WFG/NBM

One of George Frederic Matthew's specimen boxes from the collection of the Natural History Society of New Brunswick, now in the New Brunswick Museum. The box contains a specimen of the brachiopod *Camarella cambrica*, a Cambrian fossil collected in East Bay, Cape Breton. NBM

Despite the fact that he had no university education and was never associated with a university, George Matthew received many honours in recognition of his remarkable scientific career. Among them were honorary doctorates from the University of New Brunswick and Laval University; a charter membership in the Royal Society of Canada and the presidency of its geological and biological sections; and a fellowship in the Royal Geographical Society, which awarded him its Murcheson Medal.

George Mercer Dawson, May 1885. W.J. Topley/NAC PA-026689

GEORGE MERCER DAWSON

George Mercer Dawson (1849-1901), a geologist, paleontologist, and ethnographer, graduated with great distinction from the Royal School of Mines in London, receiving medals in geology, paleontology, and natural history. He was a geologist and botanist on the North American Boundary Commission, which surveyed the border between the United States and Canada, and his report is hailed as "one of the classics of Canadian geology." In 1884, with A.R.C. Selwyn, Dawson wrote the first comprehensive work on Canada's physiography, *Descriptive Sketch of the Physical Geography and Geology of the Dominion of Canada*.

Like his famous father, Sir William Dawson, George Dawson was born in Pictou, Nova Scotia. He began fieldwork in biology and geology at a very young age, collecting specimens for the McGill College Museum, which his father developed into the Peter Redpath Museum. Like his father, he became a distinguished geologist and paleontologist, with a practical eye for the development of mineral resources and scientific agriculture. Unlike his father, however, George Dawson concurred with Darwin's views of the development of life as shown in the fossil record, and while at the Royal School of Mines in London he studied with Thomas Henry Huxley, "Darwin's bulldog."

At age nine, George entered high school in Montreal; however, he contracted tuberculosis of the spine, and he didn't graduate until he was eighteen. The disease left him physically underdeveloped

George Mercer Dawson, third from the left in the back row, with members of the British North America Boundary Commission, at Dufferin, Manitoba, June 2, 1873. NAC PA-074675

and with a curvature of the spine, seemingly ill equipped for the strenuous life of a geological and geographical explorer. However, his apparent disability proved to be no handicap when he joined the Geological Survey of Canada as naturalist and geologist in 1875; he became assistant director in 1883 and director in 1895. His accomplishments raised the standards of the survey itself and thus its economic value to industry and the government. The western mandate of the GSC was "to outline geological structures, to assess their mineral wealth and agricultural potential, and to advise in the construction of a railway to the Pacific." Dawson's magnificent reports fully served all of these objectives.

While Dawson explored and mapped the geology of British Columbia, he became well acquainted with the Native people.

Intending to influence the formulation of government policy, he included information about their culturally advanced state in his reports. His ethnological observations, pioneering research, and passion for collecting Native artifacts and art attracted international attention and led people to call him "a father of Canadian anthropology." Dawson also collected — and encouraged others to collect — the vertebrate fossils, notably of dinosaurs, so plentiful in British Columbia and Alberta. Thus he laid the foundation for the collections in the Canadian Museum of Civilization and the Canadian Museum of Natural History in Ottawa.

In geology, Dawson's predictions of mineral wealth led to the recognition of the vast coal resources of Alberta and the distribution of gold in the Yukon. Indeed, his Yukon studies proved so valuable that Dawson City is named after him.

George Dawson received honours, recognition, and appointments. He was a charter member of the Royal Society of Canada, a fellow of the Canadian Mining Institute and the Geological Society of America, and a fellow and president of the Royal Society of London. Princeton, McGill, Toronto, and Queens universities conferred honorary doctorates on him, and he was made a Companion of the Order of St. Michael and St. George. In 1937, he was named a National Historic Person of Canada. His life and work set a great example of "closing the gap between scientific knowledge and practical results."

Loring Woart Bailey conducting a chemical experiment at the University of New Brunswick. PANB P37-513

LORING WOART BAILEY

Loring Woart Bailey (1839-1925), a geologist and educator, observed no boundaries among the natural sciences. The son of the first professor of chemistry and geology at West Point, he accepted the position of professor of chemistry and natural science at King's College, the forerunner of the University of New Brunswick, when he was only twenty-one years old. There, Bailey carried a full teaching and research load for forty-six years, retiring as professor emeritus in 1907 at the age of sixty-eight. He was the "shining example of the ideal man of science" and "the most prolific scientist in the University's history."

Having studied at Brown and Harvard universities under the greatest scientists of his day, including Asa Grey, Louis Agassiz, and Josiah P. Cooke, Bailey had a running start and high examples to follow. In his later years, he told his son, "I have often wondered at the temerity which led me, then a youth of scarcely twenty-one years and wholly inexperienced, to accept so important a position involving duties now distributed among not less than four professors; and that too with most insufficient preparation — some knowledge of chemistry, but little of botany and still less of zoology, and very little indeed of geology or mineralogy." And yet, this "veritable one-man department of all the non-humanities" was, according to a history of the University of New Brunswick, "destined to make by far the largest contribution to knowledge of any member of the faculty throughout the entire history of the university."

Bailey's fieldwork covered a wide area and took him on mineralogical and geological surveys of Nova Scotia, Prince Edward Island, Quebec, and Maine, as well as the length and breadth of New Brunswick. In all his work, he collaborated with local geologists. He worked especially closely with George Matthew, whose activities paralleled his own. These two remarkable scientists remained fast personal friends as well as professional colleagues throughout their lives.

Geochemistry — discovering the chemical composition of earth materials — was Bailey's specialty, but it was very much at a pioneering stage. With his impressive knowledge of chemistry, he decoded the chemical composition of the earth materials he and others collected. This analysis revealed crucial environmental information. For example, the chemistry of rocks surrounding fossil finds provided a record of conditions during the formation of the rock and the fossilization of the organism. Geochemistry could determine whether an animal lived in fresh or salt water, as well as whether that water was cold or warm.

As well as teaching science to university students, Bailey led in extension activities, providing educational opportunities for those outside the university. He took on the curator's job in the university's teaching museum, and he was a principal supporter of the Mechanics' Institutes that had sprung up in several Maritime centres. These offered intellectual pursuits to people with a passion for learning but without formal education. At the Saint John Mechanics' Institute (which, like many others, became a Natural History Society), Bailey delivered talks on a rich variety of subjects, including "Electricity and X-rays," "Geology of the St. John River Valley," "Prehistoric Man in America," "Volcanoes," "Our Soils," "Diatoms, Microscopic Algae, Rocks and What They Tell Us," and "The Body Covering of Animals."

Bailey had few colleagues in Fredericton, or even in the Maritimes, but he perceived this to be somewhat of a blessing, and it has certainly proven a blessing to posterity. He maintained a large correspondence with other leading scientists, and also he had a territory that had not been worked over by others. As he wrote a friend in 1876, "The isolation, more particularly from scientific centres and scientific co-labourers, has always been a great drawback in my position — there being not more than two or three persons in the whole province, and *none* in Fredericton, who know anything or care anything about the pursuits in which my pleasure is chiefly sought. However, I have the satisfaction of knowing that I work in a comparatively unexplored field and hope to lay a good foundation here, upon which in the future others may build."

Over the years, Dalhousie University and the University of New Brunswick conferred honorary doctorates on Loring Woart Bailey. He was a charter fellow of the Royal Society of Canada and a member of other learned societies in Canada and the United States, and he contributed many reports and papers to the Canadian Geological Survey, the Royal Society of Canada, the Natural History Society of New Brunswick, and the Fredericton Natural History Society.

Bailey wrote the fundamental books about New Brunswick geology, *Report on the Mines and Minerals of New Brunswick* and *Observations on the Geology of Southern New Brunswick*, and he contributed chapters on the geology of New Brunswick to Sir William Dawson's famous book *Acadian Geology*. His comprehensive textbook for undergraduates, *Elementary Natural History,* was published in 1887 and became a standard work that remained in use for at least forty years. As well, Bailey wrote over two hundred monographs and papers, which were published by learned societies in Canada and the United States.

Like his colleague, George Matthew, Loring Bailey was led or driven to investigate, to record, to collaborate, to advance knowledge, to teach, and to demonstrate. At the same time, he committed himself to opening the minds of everyone with whom he came in contact to the natural world around them. His admirers have left impassioned records attesting to his contributions to the earth sciences and his enthusiastic training and encouragement of others.

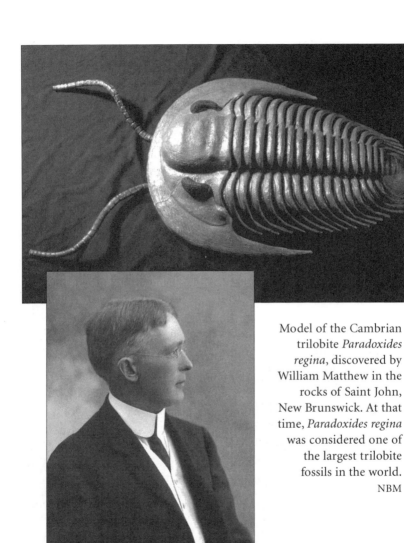

Model of the Cambrian trilobite *Paradoxides regina*, discovered by William Matthew in the rocks of Saint John, New Brunswick. At that time, *Paradoxides regina* was considered one of the largest trilobite fossils in the world.
NBM

William Diller Matthew, 1916. NBM

WILLIAM DILLER MATTHEW

William Diller Matthew (1871-1930), the celebrated vertebrate paleontologist, became known as the father of mammalian paleontology. A native of Saint John, he received early scientific training from his father, George Matthew, the noted geologist and paleontologist. He joined his father's field investigations when he was still a small child, and he graduated from high school when he was only thirteen. Because he was too young to go to university in Fredericton, he spent three years working in a law office. During that time, in the Cambrian shales near his home, he found one of the most astonishing fossils that had ever been discovered: the giant trilobite. To honour Queen Victoria, his father named this amazing creature *Paradoxides regina*.

Matthew studied geology at the University of New Brunswick under his father's friend, Dr. Loring Woart Bailey, and then did graduate work at Columbia University under the great paleontologist Henry F. Osborn. After his graduation, he went to work in the department of vertebrate paleontology at the American Museum of Natural History. He served there for over thirty years, becoming curator-in-chief of his department and gaining international fame. In 1927, he founded the department of paleontology at the University of California, Berkeley.

Matthew's training in both geology and paleontology enabled him to make one of his most significant contributions to science. He simultaneously collected fossils and stratigraphic information about the rocks surrounding them, and thus he came to recognize the important part climate has played in evolution; his most important book is probably *Climate and Evolution* (1915). His careful field investigations and penetrating research studies led to his immense intellectual contributions to the study of fossil mammals. He organized the plentiful fossilized remains of horses to demonstrate a full spectrum of evolution, a study that is still relevant today. This superb mammalian paleontologist and important biographic theorist was one of the greatest students of fossil mammals in the history of paleontology.

Matthew spent several summers in New Brunswick conducting geological surveys and collecting Cambrian fossils, and his Columbia thesis and an impressive number of his later works are on New Brunswick subjects. Altogether, Matthew published over two hundred and forty papers in publications such as the *Annals of the New York Academy of Sciences*, the Natural History Society of New Brunswick *Bulletin*, and the *Bulletin* of the Geological Society of America.

In 1992, E.H. Colbert, an eminent paleontologist and Matthew's son-in-law, said, "The work of William Diller Matthew lives on, and today his descriptions and studies of fossil mammals are widely used and respected by a generation of paleontologists more than six decades removed."

CHARLES P. ALLEN HIGH
196 ROCKY LAKE DR.
BEDFORD, N.S. B4A 2T6

CHARLES FREDERICK HARTT

Charles Frederick Hartt (1840-1878), born in Fredericton, was a naturalist with special interest in geology and paleontology. Early in his career he discovered fossil insects near Saint John, and this much celebrated accomplishment enhanced his reputation greatly, particularly after Darwin referred to it in *The Descent of Man.* With George Frederic Matthew and Loring W. Bailey (see pages 67 and 75), Hartt conducted important fieldwork in the Maritimes, but he is best known for his geological survey of Brazil and his published volume *The Geology and Physical Geography of Brazil.*

Hartt graduated from Acadia University in 1860. From 1861 to 1864, he studied at Harvard on the invitation of the great natural historian Louis Agassiz. When Agassiz organized an expedition to Brazil in 1865, he appointed Fred Hartt geologist, and the geology of Brazil became the New Brunswick scientist's greatest interest.

In 1868, Hartt taught natural history at Vassar College, but he resigned that post to accept the position of head of the geology department at Cornell University. Meanwhile, he continued to participate in expeditions to Brazil. In 1870, he organized his largest expedition, taking along another professor and a number of students from Cornell. In 1875, Hartt became chief of the Geological Commission of the Empire of Brazil, and he spent the next three years directing expeditions to various parts of the country. In 1878, his brilliant career was cut short when he died in Rio de Janeiro of yellow fever. He was only thirty-eight. George Matthew published

Charles Frederick Hartt lecturing on batrachians, c. 1895. WFG/NBM

Fossil insect wing, *Lithentomum hartii* Scudder, found by Charles Frederick Hartt in Saint John, New Brunswick, between 1860 and 1862. Named for Hartt, this is one of a series of insect fossils believed at the time to be of Devonian age and to comprise the oldest fossil insect assemblage known in the world. The fossils are now known to be from the Upper Carboniferous era, about 300 millions years old. NBM

an eloquent article in the Natural History Society of New Brunswick's *Bulletin* on that occasion, and Hartt's associate, Dr. John Casper Branner wrote, "It is not difficult to sum up Hartt's influence upon geological work in Brazil, for with very few exceptions all the work of this character which has been done in that country since 1874 is traceable, either directly or indirectly, to the impetus given it by Hartt."

WILLIAM FRANCIS GANONG

Born in West Saint John, New Brunswick, William Francis Ganong (1864-1941), botanist, cartographer, and geographer, has been called the greatest scholar ever to graduate from the University of New Brunswick. Perhaps this statement was prompted by the fact that Ganong, a botanist by profession, was above all an ardent lifelong student of his native province. Having explored all its corners, he published a series of essays, the objectives of which, he wrote, "while mainly scientific, have been also historical, especially as to facts which link man with places." Among his many academic achievements are a BA, MA, and PhD from the University of New Brunswick and a second BA from Harvard, where he served as an instructor in botany. Ganong studied for his second doctorate in Munich, returning in 1894 with a PhD in botany and an appreciation of the scientific method applied to research.

In that same year, he was appointed the first professor of botany and director of the botanic garden at Smith College, in Northampton, Massachusetts; upon his retirement in 1932, he received the distinction of professor emeritus. During his thirty-eight years at Smith, Ganong developed an outstanding botany department, he wrote four books, and he published countless articles and papers in the field of botany. In recognition of his scientific advances, the American Society of Plant Physiologists awarded him a life membership.

Fortunately, Ganong retained an intense attachment to the land-

William Francis Ganong, Rocky Brook region, York County, New Brunswick, August 14, 1909. WFG/NBM

William Francis Ganong and his wife, Anna Hobbett Ganong, on a camping trip, Bainberry Hill, New Brunswick, August 18, 1923. WFG/NBM

scape of New Brunswick and continued his study of its historical and natural geography. Growing up in St. Stephen, he had been fascinated by the flora and fauna, the prehistoric portages, and the traces of early European contact and the first permanent European settlements. He returned every summer to study the natural and human history of his native province, and the writings that resulted from his rigorous investigations reveal much about New Brunswick's past. "He was, in effect, a one-man army of scholars, single-mindedly attempting to give to New Brunswickers something of the heritage that he felt they needed to begin to build a society of the highest calibre," M.B. Caron wrote.

Ganong remains one of the most comprehensive authorities on the early maps of the northeastern part of North America. He published the first of his nine cartographical studies in the Royal Society of Canada's *Transactions* in 1887 and the last in 1937; thirty years later, the Dominion archivist observed that those papers were "still an indispensable reference, written by the chief authority on the early cartography of Canada's east coast."

The same may be said of all of Ganong's published works dealing with New Brunswick. His exhaustive research of the available literature on any topic was accompanied by field investigation. He travelled to every corner of the province, by walking and canoeing and by horse-drawn wagon and his Model T Ford. He applied the scientific method to his studies in natural history and geography, and his cultural investigations included many of the techniques employed by modern day oral historians.

Ganong's essays appeared in a wonderful variety of publications, both learned and popular. He followed his investigation of the development of coastal peat bogs by a detailed study of the vegetation of the salt and diked marshes at the head of the Bay of Fundy; this work has become a classic reference for North American ecologists. His library and writings, archived in the New Brunswick Museum, continue to be an invaluable source of information, consulted regularly by researchers.

William Ganong was awarded life membership in the Natural History Society of New Brunswick, and he received the Royal Society of Canada's Tyrell Medal and various honorary degrees. He remained an active director of Ganong Brothers, the candy business which his father and uncle established in St. Stephen. In 1945, he was named a National Historic Person of Canada. A tribute published in the Royal Society of Canada's *Transactions* says, "His one ambition was to do his work in the most thorough manner with complete honesty of purpose as his guiding principle."

HENRY G.C. KETCHUM

Henry G.C. Ketchum (1839-1896) was the University of New Brunswick's first graduate in civil engineering. Born in Fredericton, he came to be called New Brunswick's most eminent engineer, and he developed into one of Canada's greatest engineers. Even before he graduated, he worked during the summers on the European and North American Railway line that would connect Halifax with Bangor, Maine, rising to the position of assistant construction engineer on the Saint John to Shediac section. In 1860, working as an construction engineer for the São Paulo Railway in Brazil, he supervised the building of the twelve-span Megy Viaduct, which was borne on 180-foot iron columns. This great feat of engineering solidified his international reputation for skill and boldness.

After a study tour in England and his election to the British Institute of Civil Engineers, Ketchum returned to New Brunswick and resumed his career in railway construction, and here he found both challenges and triumphs. Once again he worked for the European and North American Railway, this time on a line between Moncton and Truro. This project ran into many difficulties, largely political, but while surveying the part of the line that would cross the Isthmus of Chignecto, Ketchum figured out how to build the Chignecto Marine Transport Railway. He envisioned transporting ships overland between the Bay of Fundy and Northumberland Strait; thus they would not have to sail all the way around Nova Scotia, and the sea distance between Quebec and the Bay of Fundy

An artist's conception of a sailing vessel being transported on the Chignecto Marine Transport Railway. UNB/HIL

would be more than a thousand miles shorter. Plans for a canal had been made at various times, but they had been abandoned because they were too expensive. Ketchum's ship railway would lift ships out of the water, transport them across the isthmus by rail, and relaunch them on the other side.

Ketchum told the World's Columbian Water Congress in Chicago, "The transportation of heavy and bulky merchandise over great distances at a cheap cost is of vital importance to consumers everywhere, and this is best done by water. Water carriage has the advantage over railways that railways have over common roads. Ship railway transportation combined with water carriage, by

Henry G.C. Ketchum, the University of New Brunswick's first graduate in civil engineering. UNB/HIL

A drawing showing how ships would be lifted from the water, transported, and lowered back into the water. UNB/HIL

avoiding transhipment of freight, by short cuts over isthmuses, by the saving of distance, and by avoiding the dangers of the sea, has a manifest advantage over common railways. The introduction of ship railways will mark a revolution in means of transport."

Prominent British and American engineers endorsed Ketchum's plans. He did the necessary design work, attracted English financiers, and gained government support, and construction began in the fall of 1888. Most of the work in laying the double set of tracks was done by a work force of four thousand men wielding picks and shovels. The road bed was completely straight and mostly level, and it was designed to withstand the strain of carrying ships with a displacement of up to two thousand tons, a maximum length of 235 feet, and a width of fifty-six feet. Unfortunately, after three years of building and less than a year before the project's completion, the principal banker suffered a crisis and withdrew its support. Ketchum looked for new backers, but he was unsuccessful, and despite controversy in parliament, the government also ceased to provide funding. Ketchum's magnificent dream was never realized, and one of the greatest engineering projects ever undertaken in Canada came to naught. Had it been completed, this unique ship railway would have been one of the engineering wonders of the world.

Ketchum's enthusiasm for his great idea continued to the end of his life, and he was buried — as he wished — near the marine railway line. Today, the site of the Chignecto Marine Transport Railway is a national historic site, "a straight line of embankments and cuttings, stretching more than seventeen miles across the Isthmus of Chignecto from Cumberland Basin to Baie Verte." It was said of Henry Ketchum that he lived to build. He initiated and developed some of New Brunswick's most important railway lines, and his ability to express daring vision through responsible construction has inspired the generations of civil engineers who have followed in his footsteps.

GRACE ANNIE LOCKHART

Born in Saint John, New Brunswick, Grace Annie Lockhart (1855–1916) was a bright young woman of sixteen when she enrolled at Sackville Ladies' Academy, intent on studying science. However, the only diploma available to her at that time was Mistress of Liberal Arts, after a two-year program. Lockhart graduated in 1874, but, determined to take science courses, she registered at Mount Allison University. On May 25, 1875, Grace Annie Lockhart became a Bachelor of Science and English Literature, the first woman in the British Empire to obtain a university degree of any kind. Not only was she the only female to graduate, she was, in fact, the only female student on the campus. Her achievement marked a great milestone in the history of Canadian women and Canadian education.

In 1855, the year of Lockhart's birth, Mount Allison University had issued official statements to the effect that "the introduction of abstruser sciences into a course of study for females is of the highest utility" and "the University has received indisputable evidence that female students are as capable as males." Mount Allison's approach to education was strongly influenced by its sister university, Wesleyan College, in Connecticut. In the United States, more weight had been placed on female education, particularly by the evangelical denominations, since it was believed that as mothers, women would be responsible for the moral conduct of their children. Still, women remained blocked from science programs for many years. In fact, when Lockhart entered Sackville Ladies'

Mount Allison University graduating class of 1875, including Grace Anne Lockhart. MA f20 #005 041

Academy, the sister institution to Mount Allison University, sixteen years after the university's bold statements concerning women's aptitudes and abilities, the only diploma program available to her was in arts, including subjects such as music, drawing, painting, romance languages, and other fields of study considered appropriate training in social graces. Like many universities, Mount Allison had opened its doors to women but limited their academic program.

Still, Lockhart persevered and took her science degree in 1875. By contrast, Oxford's examinations had been open to women from the 1870s onward, but degrees were withheld from them until 1920.

For her landmark accomplishment, the Historic Sites and Monuments Board named Grace Annie Lockhart a National Historic Person of Canada in 1991.

WILLIAM MacINTOSH

Born in Edinburgh, William MacIntosh (1867-1950) and his family immigrated to Canada in 1869 and settled in Saint John, where young William grew up. By training an architect and landscape gardener, he was most interested in the natural world. Membership in the Natural History Society built upon this enthusiasm. MacIntosh became the honorary curator of the society's museum in 1898 and its first full-time curator in 1907.

When the New Brunswick Museum was formed in 1930, MacIntosh took on the position of director. It was his task to incorporate into a single permanent collection the many items that the new museum had acquired from the Gesner, Mechanics' Institute and Natural History Society museums. At the same time, MacIntosh served as provincial entomologist, and he taught summer courses in many subjects related to the natural world. These courses included instruction about animal life and natural history, exploration of the St. John River and its tributaries by canoe, and the study of prehistoric Indian campsites. In addition, MacIntosh regularly contributed articles to the Natural History Society's *Bulletin*.

In 1934, MacIntosh received an honorary Doctor of Science degree from the University of New Brunswick in recognition of his outstanding contributions in the fields of anthropology, museology, and entomology. In that same year, the Carnegie Corporation

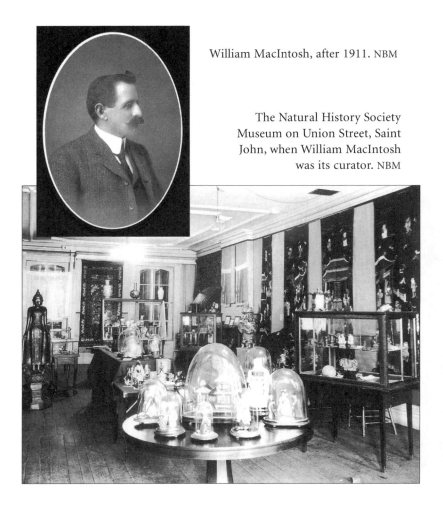

William MacIntosh, after 1911. NBM

The Natural History Society Museum on Union Street, Saint John, when William MacIntosh was its curator. NBM

awarded him a special grant for his educational activities, particularly for his highly effective use of duplicate museum specimens in a lending service that he made available to every school in New Brunswick.

JOHN CLARENCE WEBSTER

The outstanding career of John Clarence Webster (1863-1950) spanned two continents, two major disciplines, and took place during two centuries. It included the study and practice of medicine, surgery, scientific research, writing, and publishing in Europe and in North America. His career also included the study of Canadian history, a subject about which he published many books and papers.

Webster was born and grew up in Shediac, New Brunswick. At fifteen, he left home to attend Mount Allison University, and he graduated with a Bachelor of Arts degree in 1882. He then studied at Edinburgh University in Scotland, receiving his degree in medicine in 1891. Webster proved to be an outstanding student, receiving a total of fourteen prizes and scholarships during his years at Edinburgh University. He won a gold medal for his thesis, which the university published and distributed widely. After working for a time in continental Europe, he returned to Scotland; in 1891-1892, he was lecturer in obstetrics and gynecology, and then, until 1896, he was assistant professor in the Department of Midwifery and the Diseases of Women at the University of Edinburgh.

Returning to North America in 1896, Webster first settled in Montreal and took the positions of assistant gynecologist at the Royal Victoria Hospital and lecturer in gynecology at McGill University. In 1899, he left to take up an appointment as professor and head of obstetrics and gynecology at the Rush Medical School in Chicago. At Rush, Webster found a rich clinical field for his talents and keen intellect. He continued his investigations in obstetrics and gynecology there, earning international fame for books and papers sharing his groundbreaking research into subjects including ectopic pregnancy, the human placenta, and the anatomy of pregnancy.

In 1920, Webster left medicine, although at fifty-seven he was at the peak of his career, and retired to live in Shediac once more. Here, he threw himself into the study of history, soon narrowing his focus to the history of his home region. Beginning with research on the Acadians, he published and lectured on historic sites and on many aspects of New Brunswick history.

Webster acted as the driving force behind the restorations of Fort Beauséjour and Fortress Louisbourg. In 1923, he became a member of the Historic Sites and Monuments Board, and he chaired the board from 1943 until he died. He and his wife, Alice Lusk Webster, an art historian and translator, donated large amounts of money and their very considerable collections of art and artifacts to the New Brunswick Museum, and they also donated much time to the museum's establishment and operation.

John Clarence Webster earned many honours during his distinguished career. He was named honorary fellow of the Edinburgh obstetrical society, and he was a fellow of the Royal Society of Edinburgh, the Royal College of Physicians, the American College of Surgeons, and the American Association for the Advancement of Science. The University of Illinois named him professor emeritus of obstetrics and gynecology in 1942, and King George V made him a Companion of the Order of St. Michael and St. George. In 1950, he was named a National Historic Person of Canada.

John Clarence Webster, by James Cadenhead, 1896. NBM

John Clarence Webster in the library of his home in Shediac, New Brunswick, 1931. NBM

Alexander Graham Bell. NAC C-008355

Top: The Silver Dart in flight. NAC PA-061741

ALEXANDER GRAHAM BELL, MABEL HUBBARD BELL, J.A.D. McCURDY, and FREDERICK "CASEY" BALDWIN

The many achievements of Alexander Graham Bell (1847-1922) are well documented, but only by working with others, especially J.A.D. McCurdy (1886-1961), Frederick "Casey" Baldwin (1882-1948), and Mabel Hubbard Bell (1857-1923), was he able to participate in significant developments in mechanically powered transport, both in the air and on the water. The greatest of their feats was the making of "a practical flying machine driven through the air by its own power and carrying a man." Their second groundbreaking invention was the hydrofoil. The three men worked as a team, the Aerial Experiment Association, which was organized and financed by Mabel Hubbard Bell, Alexander's wife; the Aerial Experiment Association was the first organization for scientific research to be founded and supported by a woman.

Alexander Graham Bell was born in Scotland and immigrated to Ontario with his parents. His father was a teacher of the deaf, and, in 1871, Alexander went to the School of Oratory in Boston as a substitute for his father. There, he met and married one of the students, the intelligent, wealthy, and well-connected Mabel Gardiner Hubbard. An exceptional woman, she collaborated with him in many ways. His invention of the telephone, a spinoff from his efforts to enable the deaf to communicate, had made Bell famous, and he was already a worldwide celebrity when, in 1885, he and Mabel discovered the beauty of Baddeck, Cape Breton, and the Bras d'Or lakes. There they built a spacious estate they named

Beinn Bhreagh (pronounced "ben vreeah"), Gaelic for "beautiful mountain." For the rest of his life, they spent the summer months there. Bell set up laboratories and conducted groundbreaking experiments at Beinn Bhreagh, including those with kites that led to powered flight.

Powered, manned flight was an early interest of Bell's. After Samuel Pierpont Langley's experiments with mechanical flight in the United States, and independent of the Wright brothers, Bell had become deeply involved in trials with tetrahedral kites that could carry a human. As Buckminster Fuller has pointed out, Bell's work was some fifteen years ahead of anybody else's. By the early years of the twentieth century, the achievement of manned heavier-than-air flight was just a matter of time. Experiments were underway in the United States and France. Wallace Turnbull, working out principles governing lift and thrust in wing sections and studying propeller shapes in Rothesay, New Brunswick, kept in touch with Bell's work with kites. All were convinced that solutions were close.

At the same time, Glenn Curtiss was building lightweight engines an Hammondsport, in upstate New York, and with fellow American, Lieutenant Thomas Selfridge, he was keenly interested in flight. Frederick "Casey" Baldwin, grandson of the great Robert Baldwin, was about to graduate in mechanical and electrical engineering from the University of Toronto, as was his close friend, Douglas McCurdy, of Baddeck, and both men were thinking about flight. Mabel Bell encouraged them all to come to Beinn Bhreagh, challenging them to join in the excitement of discovery. The stage was set.

Bell, Curtiss, Selfridge, Baldwin, and McCurdy participated together in the design and building of a functional aircraft. United in a great adventure, they were known as "A.E.A. boys." They made their first flights at Hammondsport; on March 12, 1908, Casey Baldwin, in *Red Wing*, was the first British subject to fly. Indeed,

The Aerial Experiment Association: Glenn Curtiss, Douglas McCurdy, Alexander Graham Bell, Casey Baldwin, and Thomas Selfridge. PC/AGB

since the Wrights had been so secretive, a claim was made that it was Baldwin who gave "the first public exhibition of a heavier-than-air machine in America." Less than a year later, on February 23, 1909, Douglas McCurdy took the *Silver Dart* off from the ice at Baddeck and flew for half a mile, the first flight in the British Empire. All of the equipment, such as the tricycle landing gear and ailerons, were products of their Aerial Experiment Association, and McCurdy was its chief designer and builder. The very next day, he flew a distance of four miles.

Having achieved this success, the Aerial Experiment Association dissolved. J.A. Wilson, writing in the *Journal of the Engineering Institute of Canada* in 1937, says, "The Association's life was short but brilliant. No contemporary work was more successful or productive of lasting results."

McCurdy was issued Canada's first pilot's license in 1910, he sent the first wireless messages from an aircraft, and he set the word's first record for over-water flight in 1911. In 1914, he established a flying school in Toronto for the training of war pilots. He then founded Aircraft Industries Ltd.; in 1929, the company merged with a company founded by Curtiss. McCurdy then turned to public service, becoming Canada's assistant director general of aircraft production in the Second World War. After returning to Baddeck, he was appointed lieutenant-governor of Nova Scotia, a post he held from 1947 until 1952. He was named a National Historic Person of Canada in 1974.

Casey Baldwin remained in Baddeck. Becoming as close as a son to the Bells, he lived with them at Beinn Bhreagh, and when he married, he and his wife moved into another house on the property. He managed both the estate and the laboratory, and he and Bell continued with many projects. They realized that large aircraft would need to taxi for a long distance to get up enough speed to achieve lift off, and they, like others, became convinced that this would be possible only if the aircraft could take off from water. Investigations and experiments with pontoons and hydrofoils were underway in the United States, England, and Italy, so the Bells and Baldwins travelled to Italy. There they purchased patents that gave them a head start in the development of the hydrofoil.

Progress in aircraft design meant that airports became quite capable of accommodating large planes, and the use of aircraft in World War I drove tremendous development in many aspects of flight technology. Therefore, Bell and Baldwin began to work on hydrofoil boats, especially for the pursuit of submarines. They failed to persuade naval authorities in the United States and Canada to back their venture, but they persisted in developing the hydrofoil through several versions. The last one was the HD-4. In 1919, fitted with two 350-horsepower Liberty aircraft engines provided by the American Navy, the HD-4 easily set a water speed record of 70.86 miles per hour, a record that held for ten years. A full-sized replica of the HD-4 is now on display in the Bell Museum at Baddeck.

Bell and Baldwin's hydrofoil, the HD-4, 1919. PC/AGB

Alexander Graham Bell was a discoverer and inventor until his death in 1922; in his lifetime he himself was granted nineteen patents, and he shared eleven with his collaborators. In the twenty years after Bell's death, Casey Baldwin continued to make startling advances in high-speed water transportation at Baddeck. Over the years, his health failed gradually, and when he died in 1948, he was laid to rest near the Bells at their beloved Beinn Bhreagh.

Bell's "Message to Children" articulates his principles of invention: "Don't keep forever on the public road, going only where others have gone. Leave the beaten track occasionally and dive into the woods. You will be certain to find something you have never seen before. Of course, it will be a little thing, but do not ignore it; follow it up, explore all around it; one discovery will lead to another, and before you know it, you will have something worth thinking about to occupy your mind. All really big discoveries are the result of thought."

Wallace Rupert Turnbull stands in front of an Avro 504 airplane equipped with Turnbull's electrically operated variable pitch propeller, Camp Borden, 1927. CAM

WALLACE RUPERT TURNBULL

Wallace Rupert Turnbull (1870-1954), a pioneering aeronautical engineer, was one of those fortunate individuals for whom most things in life go smoothly. Born to a prosperous family in Saint John, he had a privileged upbringing and studied mechanical and electrical engineering, first at Cornell University and then in Berlin and Heidelberg. He worked in industry for several years, and then — judging that manned flight was close at hand — he established an aeronautical laboratory at his home in Rothesay, New Brunswick. As a result of his invention of the variable-pitch propeller, fame and fortune came to him, yet his contributions to the science of flight both precede and go far beyond that single great achievement.

Before anyone had actually flown, Turnbull recognized the importance of scientific knowledge in aviation. He said, "I am most proud of my achievement in bringing science to aeronautics: of showing the world that only through an understanding of the fundamental principles governing flight — such as lift, drag, drift, and centre of pressure — could we hope to build stable, efficient, and safe aircraft." He continued, "To advance my research, I needed a constant supply of controlled air currents, so I built the first wind tunnel in Canada, in Anderson's barn. In it, I improved wing designs and discovered the propeller that made aviation and air transport a commercial possibility." His inventions also included a propeller-testing vehicle, the "Wind Wagon," which ran on railway tracks. In 1902, Turnbull began devising the technology that would help make powered, manned flight a reality, and he didn't stop until the end of his life.

Canada's first aeronautical engineer, Turnbull wrote extensively and contributed to a number of learned journals. Collaborating and freely sharing his research results, he eagerly added to the growing body of knowledge of the emerging phenomenon of flight by humans. In 1909, the Royal Aeronautical Society awarded him a medal for his paper, "The Efficiency of Aeroplanes, Propellers, Motors, etc.," published in 1908 in its *Aeronautical Journal*.

When World War I broke out, Turnbull made his way to England and volunteered to serve without pay. He was assigned to an aircraft manufacturing company as a designer of aircraft and placed in charge of the development section; there he could use his talents in developing original ideas. His inventiveness continued unabated, and the long list of his successes included new propeller designs, improved seaplane floats, screens to protect ships from torpedoes, a "bomb sighter for aeroplanes," and consolidated wood, a product of very high strength. Using consolidated wood and following Turnbull's design, thousands of propellers were produced for Allied warplanes.

Toward the end of the war, Turnbull turned his mind to the variable-pitch propeller. Back in Rothesay, his experiments had determined the most efficient angle at which a blade cuts the air, and he now faced the challenge of how to change a propeller's pitch while the aircraft was in flight. By 1927 he had succeeded, and the controllable propeller was a demonstrated reality. Although he sold his patent rights two years later, he continued to receive royalties for this invention.

For more than fifty years, Turnbull's fertile mind investigated a wide range of problems in the field of aeronautics. As well, he built aircraft engines and hydroplane boats, became involved in timber science, and patented an engine to harness tidal and wave power.

Though unknown to the general public, Turnbull was very well known among the informed and made his findings freely available to others of like interests. He shared his work with Alexander Graham Bell and the Aerial Experiment Association in their aviation experiments in Baddeck, and he was committed to flight before anyone anywhere had flown a heavier-than-air craft. "I knew even as early as 1900," he said, "that the airplane was on its way and soon flying machines would be a reality. I also knew that before it could happen there were hundreds of practical problems that required a creative, yet methodical mind, such as mine."

An active member of the National Research Council and one of the select few to be elected to the Canadian Science and Engineering Hall of Fame, he was named an honorary fellow of the Canadian Aeronautical Institute, its highest honour, and in 1960 he was declared a National Historic Person of Canada. Today, Wallace Turnbull is honoured at the National Museum of Science and Technology, and his variable-pitch propeller — which made commercial flight economical and practical — is a feature in the permanent collection of Canada's National Aviation Museum.

Georgina Fane Pope. Her Royal Red Cross medal
is attached to her uniform above her other medals.
PARO-PEI Camera Club Coll. 2320/103.3

GEORGINA FANE POPE

Charlottetown native Georgina Fane Pope (1862-1938) followed her ambition and became a renowned military nurse, a pioneer in providing effective and compassionate medical care to soldiers wounded in battle. Her example, her training programs, and her application of sound standards helped nursing become a respected profession within Canada's military forces.

Georgina Pope was the daughter of William Pope, one of Canada's Fathers of Confederation. Attractive and well-born, she seemed destined for the life of a provincial socialite. Instead of settling into marriage and Prince Edward Island society, however, she took the bold step of choosing a serious career in nursing. To pursue her dream, she moved to New York to study at the Bellevue Hospital School of Nursing. After she graduated, she held administrative positions at hospitals in the United States until 1899. When the Boer War broke out in South Africa, Pope volunteered her services to the British Army, and in the fall of 1899, she sailed from Quebec as the head of four nurses, Canada's first group of nursing sisters to support the British army.

For more than a year, Pope cared for the wounded in South Africa, often experiencing extreme physical and emotional hardship herself. Despite difficult conditions, she remained dedicated to modern medical standards at all times. In the process, she became a leader in maintaining hospital cleanliness, good personal hygiene,

Minnie Affleck, Nursing Sister, 1st Canadian Contingent, with a group of wounded soldiers, 1900. Affleck was one of the four nurses who went with Pope to serve in the Boer War. NAC C-051799

and good nutrition for wounded soldiers while establishing the highest possible quality of medical care in battlefield conditions.

Pope returned briefly to Canada, but in 1902, she went back to South Africa as senior sister in charge of eight nurses. By then, she and those who served under her, including Margaret Macdonald, had officially become the Canadian Army Nursing Service. In 1903, Pope received the Royal Red Cross for conspicuous service, the first Canadian ever honoured with this medal. After the British troops left South Africa, Pope served with the Canadian army reserves, and then she accepted an invitation to join the permanent forces. In 1908, she was promoted to Matron of the Canadian Army Medical Corps, and she established the uniform for the service, a distinctive blue instead of the usual army khaki.

Despite her age — she was then fifty-five — Nursing Matron

Pope went overseas in 1917 and served in both the United Kingdom and in France. In 1918, she arrived at the army hospital in Outreau just in time for a major assault. In addition to the constant stream of wounded and dying soldiers, she endured the thunder of artillery fire all around and the ever-present fear of being overrun by the enemy. Still she continued to perform her duties. At the end of June, Matron Pope was diagnosed with shell shock. Unable to serve any longer and unfit for further duty due to her "war related injuries," she returned to Canada. At that time, her colleague, Margaret Macdonald, said of her, "Whilst her individual work has always commanded the highest praise and admiration, it must also be remembered that to her is largely due the success that has ever attended the CAMC Nursing Service — a success that might never have been attained but for the high standard and lofty ideals set and maintained by Matron Pope from the inception of the Corps." In 1919, officially retired, Pope returned to Charlottetown, where she spent the rest of her life. When she died in 1938, "the Island's Florence Nightingale" lay in state at Government House before being buried with full military honours. In 1983, she was designated a National Historic Person of Canada.

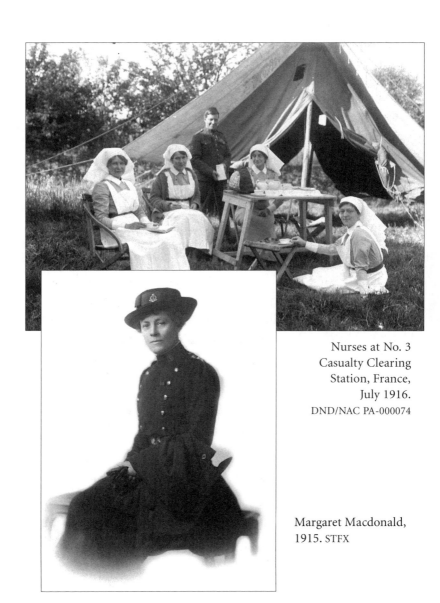

Nurses at No. 3 Casualty Clearing Station, France, July 1916.
DND/NAC PA-000074

Margaret Macdonald, 1915. STFX

MARGARET MACDONALD

Margaret Macdonald (1873-1948), born in Pictou County, Nova Scotia, was the first woman in the British Empire to hold the rank of major. She began her training as a nurse at Mount Saint Vincent College, Halifax, and continued at the New York City Training School. Upon completing her studies, she served in the Spanish American War in 1898, working in military hospitals and aboard the hospital ship *Relief*.

Although the services of female nurses had been used by the military in Canada during the North West Rebellion of 1885, they were volunteers only; their dedicated and often dangerous work had no official standing, nor did they have defined duties and rights. The Canadian Army Nursing Service was officially organized in 1901, when Canadian soldiers went to South Africa to fight in the Boer War. Volunteering for duty, Macdonald became one of the first seven members of the service.

When the Boer War ended, Macdonald insisted on being sent to Britain to study military nursing organization and practice. At the outbreak of the First World War, she was appointed Matron in Chief, and in November 1914, she received the military rank of major. Not only was she the first female British subject to hold such a rank, but Canada's nurses were unique in serving as part of the armed forces, with military ranks and a military chain of command. In Britain, Macdonald directed a huge organization of nurses stationed in hospitals and aboard hospital ships and trains, and at the same time she was responsible for nursing services and the supply of nurses in Canada. When war was declared in August 1914, there were a mere five army nursing sisters; at the time of the 1918 Armistice, a total of 3,141 nursing sisters had enlisted in the Canadian Army Medical Corps, and 2,504 of them had served overseas. These had acquired the nickname "Bluebirds" because of their bright blue cotton hospital uniforms.

Before she left active military service in 1923, Macdonald received many honours. During the war, King George V personally presented her with the Royal Red Cross, and by war's end she had also been awarded the rare Florence Nightingale Medal. The National Council of Women of Canada made her an honorary life member, and in 1920, when the Overseas Nursing Sisters Association was established, she was elected honorary president. Later, St. Francis Xavier University awarded her an honorary DDL.

When Macdonald died in 1948, she was buried at Bailey's Brook, Pictou County, Nova Scotia, the community where she was born. After her death, the Canadian Army honoured her memory by opening Margaret Macdonald House at Camp Borden, Ontario. In 1982, Margaret Macdonald was named a National Historic Person of Canada.

ROBIE TUFTS

The great Nova Scotia ornithologist and conservationist Robie Tufts (1884-1982) was born in Wolfville, Nova Scotia and lived there all his life. He spent at least ninety of his ninety-eight years vigorously influencing all around him, and that influence continues to be felt today.

Tufts could not have escaped a life concerned with natural history. His father was a professor of natural science at Acadia University, and his mother, an accomplished botanist, encouraged his early field forages for birds. As a young man, Tufts was an avid hunter, but one who would take a wounded bird home to be nursed to health and, if possible, released.

Initially choosing finance, Tufts became a banker with an assured career ahead of him. However, he could not help but observe that overhunting had drastically decreased the bird population, and he was dismayed at the general lack of conservation standards. Two years after the passage of the Migratory Birds Convention Act in 1917, he became chief migratory birds officer for the Maritime Provinces.

Tufts enforced the new laws vigorously; by 1920, he had 679 charges and convictions to his credit. He enforced the law mercilessly, though some of the offenders were his friends, and on more than one occasion Tufts paid the convicted man's fine. His devotion to conservation made converts of his captives; once, it is said, he caught two boys shooting birds in an Annapolis Valley orchard.

Robie Tufts. AU

After scolding them roundly, he invited them to his home, and there he gave them a gentler lesson in the principles of conservation. Both eventually pursued wildlife conservation as a career, one of them in a senior position at the National Museum of Natural Science in Ottawa.

Tufts had a special relationship with the Nova Scotia Museum and its director, Don Crowdis. He sold his superb egg collection to the museum, and Crowdis strenuously encouraged Tufts to write his groundbreaking book *Birds of Nova Scotia*. Tufts delayed the project for a few years, being very busy in the field, but when the museum advanced him one quarter payment for a yet undelivered one quarter of the text, his embarrassment forced him one night to "lock the door and sit down at the desk." His book, the first full volume published by the Nova Scotia Museum, was a huge success and has gone into several printings.

For many years, the Nova Scotia Museum sponsored a series of Audubon Screen Tours, which were presented to sold-out houses in a thousand-seat auditorium. The Society for the Birds of Nova Scotia was formed at the inaugural performance, with Robie Tufts as unanimous choice for president.

Tufts wrote more than 65 articles and books. Both *Nova Scotia Birds of Prey* and *Birds of Nova Scotia* remain authoritative in their field, and the newspaper columns he contributed each week to the *Halifax Chronicle Herald* were collected into the book *Birds and Their Ways*. Tufts received honorary degrees from Acadia and Dalhousie universities, and both *Asio otus tuftsi*, a subspecies of the long-eared owl, and the ornithology laboratory at Acadia University bear his name.

ILLUSTRATION CREDITS

The following credits for illustrative material are listed in alphabetical order of the abbreviations that appear with the captions. Illustrative material on page 116, courtesy of the Biology Department, Acadia University (AU); on page 104, courtesy of Canada Aviation Museum, Ottawa (CAM); on page 16, courtesy of Heritage Resources, Saint John (HRSJ); on page 12, courtesy of Kings Landing Historical Settlement (KLHS); on pages 52 and 55, courtesy of the McGill University Archives (MU); on page 92, courtesy of Mount Allison University Archives Picture Collection (MA); on pages 70, 72, 98, 110, and 112, courtesy of the National Archives of Canada (NAC) and the Department of National Defence, National Archives of Canada (DND/NAC); on pages 10, 16, 22, 30, 34, 69, 78, 82, 84, 94, and 97, courtesy of the New Brunswick Museum (NBM), including the John Clarence Webster Canadiana Collection (JCW/NBM), the William Francis Ganong Collection (WFG/NBM); on page 40, courtesy of the New Brunswick Museum and Joan Adney Dragon (NBM/JAD); on pages 18, 20, 26, 27, 32, 48, and 58, courtesy of the Nova Scotia Archives and Records Management, Halifax, NS (NSARM); on pages 101 and 103, courtesy of Parks Canada / Alexander Graham Bell National Historic Site of Canada (PC/AGB); on pages 36 and 38, courtesy of the Prince Edward Island Museum and Heritage Foundation (PEIMHF); on page 74, courtesy of the Provincial Archives of New Brunswick (PANB); on page 108, courtesy of the Public Archives and Records Office of Prince Edward Island (PARO-PEI); on page 112, courtesy of St. Francis Xavier University Archives, Dr. Ronald St. John Macdonald Collection (STFX); on page 50 courtesy of the U.S. Naval Observatory Library (USNOL); on pages 61, 62, 64, 88, and 89, courtesy of the University of New Brunswick, Archives and Special Collections, Harriet Irving Library (UNB/HIL); and on page 46, courtesy of Victoria University Library (Toronto) (VUL). All illustrations appear by permission.

ACKNOWLEDGEMENTS

Two persons in particular provided generous help with this volume. A special salute to them is in order, for I am especially grateful for all their help. Donald K. Crowdis, former Director of the Nova Scotia Museum, showed a keen interest in this project from the start. As a dear old friend and colleague, he generously provided good counsel and popular writing ability in the scientific field, and his impressive editing skills are to be recognized on every page of the book. And I owe a huge debt of gratitude and affection to my wife, Marie MacDonald MacBeath, herself a significant achiever in the fields of chemistry and science education. She more than went the extra mile with this book-making project. I will be forever in her debt for her continued encouragement, insightful suggestions, and computer skills. It is largely thanks to her that both the exhibition and the accompanying book have come into being.

All those who have been involved in this Maritime Achievers project are most appreciative of the invaluable support of the Museum Assistance Program of Heritage Canada. Without that financial aid, the exhibition and this book would not have become a reality. The cooperation and professionalism of Goose Lane Editions, which took on the publishing of this book, is also recognized and much appreciated.

Special thanks go to New Brunswick Museum Director Jane Fullerton and curatorial staff members Randy Miller, Stephen Clayden, Peter Laroque, and Gary Hughes. Director Margaret Pacey and her Legislative Library staff could not have been more helpful. That is also true of the Provincial Archives of New Brunswick staff and that of the University of New Brunswick Library and Archives. To be mentioned as well is the assistance of Aynsley MacFarlane, of the Alexander Graham Bell National Historic Site; the National Archives of Canada; the Nova Scotia Archives; the Royal Ontario Museum; Canada Post; Kings Landing Historical Settlement; the

Prince Edward Island Public Archives, and Boyde Beck, of the Prince Edward Island Museum.

A Note on Sources

Information for *Great Maritime Achievers* has come from many sources. Encyclopedic works have been invaluable, first and foremost being the *Dictionary of Canadian Biography*, George W. Brown, David M. Hayne, Francess G. Halpenny and others, eds. (Toronto: University of Toronto Press). Others include the *Canadian Biographical Dictionary* (Toronto: Cooper, 1881); *Encyclopedia Canadiana* (Ottawa and Toronto: Grolier, 1957, 1972); James H. Marsh, ed., *The Canadian Encyclopedia* (Edmonton: Hurtig, 1985, 1988); *Dictionary of American Biography* (New York: Scribner, 1977); and *Dictionary of Scientific Biography* (New York: Scribner, 1970).

Among the books containing information on Maritime achievers are: J.J. Brown, *Ideas in Exile: A History of Canadian Invention* (Toronto: McClelland and Stewart, 1967); Alfred G. Bailey, ed., *The University of New Brunswick Memorial Volume* (Fredericton: University of New Brunswick, 1950); Frank H. Ellis, *Canada's Flying Heritage* (Toronto: University of Toronto Press, 1954); John G.Reid, *Mount Allison University: A History to 1963*. Vol. I: 1843-1914 (Toronto: University of Toronto Press, 1984); W.A. Squires, *The History and Development of the New Brunswick Museum, Saint John (1842-1945)* (Saint John: New Brunswick Museum, 1945); and G.W.L. Nicholson, *Canada's Nursing Sisters* (Toronto: Hakkert, 1975).

Some of the periodicals and serials in which these achievers made their findings known also published biographical sketches; among them are the *Transactions of the Royal Society of Canada*; the *Nova Scotia Historical Quarterly*; the Nova Scotia Institute of Natural Science *Transactions* and *Bulletin*; and the Natural History Society of New Brunswick *Bulletin*.

The agenda papers prepared for the Historic Sites and Monuments Board of Canada are a treasury of information on people who have been recommended to that body for recognition.

INDEX

A

Acadia University 81, 115, 117
Adney, Tappan 43
Aerial Experiment Association 99, 101, 107
Affleck, Minnie 110
Agassiz, Louis 75, 81
agricultural implements 36-38
agriculture 21, 41-44, 53, 61, 71
airplane 99-102
Albert County NB 12, 15
Alberta 73
Aleutian Islands, AK 47
American Association for the Advancement of Science 51, 53, 61, 96
American College of Surgeons 96
American Museum of Natural History 79
American Society of Plant Physiologists 83
anatomy 59
Annapolis Valley NS 115
anthropology 23, 73, 93
Antigonish NS 57
apples 42-43
archaeology 67

architecture 93
  marine 29
Armdale NS 25
Astronomical Society of the Pacific 51
astronomy 49, 63, 65
Atlantic Ocean 47
Atlantic Telegraph Company 47
Audubon, John James 25
Australia 35, 47
Avro 504 airplane 104

B

Baddeck NS 99-103, 107
Baie Verte NB 90
Bailey, Loring Woart 68, 74-77, 79, 81
Bailey's Brook NS 114
Bainberry Hill NB 84
Baldwin, Frederick "Casey" 99-103
Baldwin, Robert 100
Bangor, ME 87
Bay of Fundy 11, 85, 87
beacon light 17
Bedford Basin 33
Beinn Bhreagh 100, 102, 103
Bell, Alexander Graham 98-103, 107

Bell, Mabel Hubbard 99-103
Bellevue Hospital School of Nursing 109
biology 71
Boer War 109, 110, 113
botany 19, 53, 59, 75, 83, 115
Branner, John Casper 82
Bras d'Or lakes 99
Brazil 81
British Association for the Advancement of Science 53
British Columbia 72
British Institute of Civil Engineers 87
British North America Boundary Commission 72
British North America Electric Telegraph Association 45
Brogdale Horticultural Trust 43
Brown, J.J. 31
Brown University 75
Brydone Jack Astronomy Club 66

C

Cabot Strait 47
Calais ME 49
Cambridge MA 50
Camp Borden ON 104, 114
Canada Post 35
Canadian Aeronautical Institute 107
Canadian Army Medical Corps 110, 114
Canadian Army Nursing Service 110, 111, 113
Canadian Mining Institute 73
Canadian Museum of Civilization 73
Canadian Museum of Natural History 73
Canadian Science and Engineering Hall of Fame 107
Canadian Society of Civil Engineers 47
Cape Tormentine PEI 45
Cape Traverse NB 45
Carnegie Corporation 93
Caron, M.B. 85
cartography 83, 85
Charlottetown PEI 12, 109, 111
chemistry 17, 53, 59, 75, 76
Chignecto, Isthmus of 87-90
Chignecto Marine Transport Railway 87-90
Colbert, E.H. 80
Columbia University 79
Confederation 35, 39, 58, 65, 109
Connell Memorial Herbarium 59, 60
Cooke, Josiah P. 75
Cornell University 81, 105
Cornwallis NS 11
Crowdis, Don 117
Cumberland Basin 90
Curtiss, Glenn 100, 101

D

Dalhousie University 13, 77, 117
Darwin, Charles 54, 71, 81
Dawson City YT 73
Dawson, George Mercer 70-73
Dawson, John William 13, 52, 53-57, 71, 77
Downs, Andrew 25-28
Dublin, Ireland 57
Dufferin MB 72

E

East Bay NS 69
Edinburgh Obstetrical Society 96
Edinburgh University 95
Einstein, Albert 51
engine
　amphocratic steam 17
　compound marine 29, 31
engineering 14, 63, 65
　aeronautical 100-102, 105-107
　civil 63, 87, 89, 90
　electrical 100, 105
　mechanical 100, 105
England 45
entomology 93
European and North American Railway 87

F

Faraday, Michael 11
Fenerty, Charles 32-35
First World War, *See* World War I
foghorn, steam 15-17
Fort Beauséjour 96
Forteau NL 24
Fortress Louisbourg 96
Foulis, Robert 15-17
France 111
Fredericton Athenaeum 63
Fredericton Natural History Society 77
Fredericton NB 42, 43, 59, 60, 63, 65, 76, 77, 79, 81, 87
Fuller, Buckminster 100

G

Ganong, Anna Hobbett 84
Ganong, William Francis 68, 83-86
geochemistry 76
geography 83, 85
Geological Society of America 58, 73, 80
Geological Society of London 53
Geological Survey of Canada 13, 58, 67, 72, 77
geology 12, 14, 53, 55, 57, 58, 59, 68, 71, 72, 73, 75, 77, 79, 81
George V, King 96, 114
Gesner, Abraham 10, 11-14, 53, 61
Gesner Museum 68, 93
Gisborne, Frederick Newton 45-47
Grey, Asa 75
gynecology 95-96

H

Halifax NS 13, 19, 21, 25, 28, 33, 39, 87, 113
Hall, Thomas 37-39
Hallock, Charles 28
Hamilton, Charles 34, 35
Hardy, Campbell 25, 28
Hartt, Charles Frederick 68, 81-82
Harvard University 50, 54, 75, 81, 83
Henry, Joseph 50
Historic Sites and Monuments Board 31, 66, 92, 96
Honeyman, David 57-58
Howe, Joseph 21
Huxley, Thomas Henry 71
hydrofoil 99, 102, 103

I

ichthyology 24
Iowa State College 43
Italy 102

J

Jack, William Brydone 61, 62-66
Johns Hopkins University 50
Johnston, J.F.W. 61

K

Keller, Frederich 35
Kennedy, J.E. 65
kerosene 11-12
Kerosene Gas Light Company 13
Ketchum, Henry G.C. 87-90
King's College, *See* University of New Brunswick
King's Landing Historical Settlement 44

L

Langley, Samuel Pierpoint 100
Laval University 69
Lockhart, Grace Annie 91-92
Logan, William 13, 58
London, England 57
Lorne, Marquess of 54
Lyell, Charles 13, 53, 61

M

Macdonald, Margaret 110-114
MacIntosh, William 93-94
Margaret Macdonald House 114
mathematics 49, 50, 63, 65
Matthew, George Frederic 67-69, 76, 77, 79, 81
Matthew, William Diller 68, 78-80
Maugerville NB 23
McCurdy, J.A.D. 99-103
McGill University 53, 71, 73, 95
Mechanics' Institute, Halifax 21, 27, 58, 76
Mechanics' Institute, Saint John 17, 24, 76, 93
medicine 11, 15, 59, 95, 96
Megy Viaduct 87
Migratory Birds Convention Act 115
Miles, Frederick H.C. 16
Mill Cove NB 29
mineralogy 59, 75
Mining Association of Newfoundland 47
Mispec NB 34
Moncton NB 49, 87
Montreal QC 95
Morris, Maria Frances Ann 21
Mount Allison University 91, 92, 95
Mount Saint Vincent College 113
museology 93
Museum of Comparative Zoology 54

N

National Aviation Museum 107
National Council of Women of Canada 114
National Historic Persons of Canada 14, 55, 73, 86, 92, 96, 102, 107, 111, 114
National Museum of Natural Science 117
National Museum of Science and Technology 107
National Research Council 107
natural history 13, 21, 23, 24, 25, 54, 55, 58, 59, 71, 81, 85, 93, 115
Natural History Society of New Brunswick 24, 69

New Brunswick Board of Agriculture 59
New Brunswick Legislative Assembly 17, 31
New Brunswick Museum 11, 67, 68, 69, 86, 93, 96
New Brunswick Natural History Society 68, 76, 77, 80, 82, 86, 93, 94
New Brunswick Society for the Encouragement of Agriculture, Home Manufactures, and Commerce 61
New York City Training School 113
New York NY 25, 28
Newcomb, Simon 49-51
Nightingale, Florence 111
North American Boundary Commission 71
North West Rebellion 113
Northumberland Strait 45, 87
Nova Scotia Historical Society 35
Nova Scotia Institute of Natural Science 28, 58
Nova Scotia Museum 117
Nova Scotia Telegraph Company 45
nursing, military 109, 111, 113, 114

O

obstetrics 95, 96
Ontario 13, 39
Order of St. Michael and St. George 73, 96
ornithology 115
Osborn, Henry F. 79

Overseas Nursing Sisters Association 114
Oxford University 92

P

Pacific Ocean 47
paleontology 53, 67, 71, 79, 80, 81
paper, wood-pulp 33-35
Paris, France 57
Parrsboro NS 11
Partridge Island NB 15, 16
Pennsylvania 13
Perley, Moses Henry 22, 23-24
petroleum science 14
Philadelphia, PA 57
Phillipson, D.J.C. 54
physiology 59
Pickett, Sarah Jane 30
Pictou County NS 113, 114
Pictou NS 53, 71
Pope, Georgina Fane 108-111
Pope, William 109
Princeton University 54, 73
propeller, variable-pitch 105-107
Provincial Museum of Nova Scotia 58

Q

Quebec 31, 87
Queens County NB 29
Queen's University 73

R

Reciprocity Treaty 24
Red Wing 100
Redpath, Peter 54, 71
Reindeer 30
Robb, James 60-61
Rockefeller, John D. 13

Rocky Brook NB 84
Rothesay NB 100, 105, 106
Royal Aeronautical Society 106
Royal College of Physicians 96
Royal Geographical Society 69
Royal Institution 11
Royal School of Mines 71
Royal Society of Canada 28, 47, 54, 55, 58, 69, 73, 77, 85, 86
Royal Society of Edinburgh 96
Royal Society of London 51, 54, 73
Royal Victoria Hospital 95
Rush Medical School 96

S

Sackville Ladies Academy 91, 92
St. Andrews University 63
St. Francis Xavier University 114
Saint John NB 11, 15, 17, 23, 42, 65, 67, 78, 79, 81, 82, 83, 87, 91, 93, 94, 105
Saint John Pulp and Paper Company 34
St. John River 15, 30, 93
St. John's NL 47
St. Stephen NB 85, 86
School of Oratory 99
Scotland 15, 57, 59, 63, 93, 99
Second World War, *See* World War II
Selfridge, Thomas 100, 101
Selwyn, A.R.C. 71
Sharp, Francis Peabody 40-44
Shediac NB 87, 95, 96, 97
Silver Dart 98, 101
Smith College 83
Smith, Titus 18-21, 33
Smithsonian Institution 28, 50, 54

Society for the Birds of Nova Scotia 117
South Africa 109, 110, 113
Spanish American War 113
Summerside PEI 37
surveying 13, 19, 65
Sussex NB 42

T

taxidermy 25
Telegraph and Signal Service 47
telegraphy underwater 45, 47
telephone 99
Tibbets, Benjamin 29-31
trilobite 68, 78, 79
Truro NS 53, 87
Tufts, Robie 115-117
Tupper, Charles 58
Turnbull, Wallace Rupert 100, 105-107

U

United States Naval Observatory 50
University of California 79
University of Edinburgh 59, 95
University of Illinois 96
University of New Brunswick 59, 61, 62, 63, 64, 65, 68, 69, 74, 75, 77, 79, 83, 87, 89, 93
University of St. Andrews 57, 59
University of Toronto 73, 100

V

Vassar College 81
Victoria, Queen 79

W

Wallace Bridge NS 49
Webster, Alice Lusk 96
Webster, John Clarence 95-97
Wesleyan College 91
West Point Military Academy 75
William Brydone Jack Observatory 64
Wilmot PEI 37
Wilson, J.A. 101
wind tunnel 105
Wolfville NS 115
Woodstock NB 41, 42
Woodstock Nurseries 41
World War I 102, 106, 113
World War II 102
World's Columbian Water Congress 88
Wright, Orville 100
Wright, Wilbur 100

Y

York County NB 84
Yukon Territory 73

Z

zoological gardens 25-28
Zoological Society of London 28
zoology 53, 59, 75